Chemicals in the Environment
Assessing and Managing Risk

ISSUES IN ENVIRONMENTAL SCIENCE AND TECHNOLOGY

EDITORS:

R.E. Hester, University of York, UK
R.M. Harrison, University of Birmingham, UK

EDITORIAL ADVISORY BOARD:

Sir Geoffrey Allen, Executive Advisor to Kobe Steel Ltd, UK, **A.K. Barbour**, Specialist in Environmental Science and Regulation, UK, **N.A. Burdett**, Eastern Generation Ltd, UK, **J. Cairns, Jr**, Virginia Polytechnic Institute and State University, USA, **P.A. Chave**, Water Pollution Consultant, UK, **P. Crutzen**, Max-Planck-Institut für Chemie, Germany, **S. J. de Mora**, International Atomic Energy Agency, Monaco, **P. Doyle**, Syngenta, Switzerland, **G. Eduljee**, SITA, UK, **M.J. Gittins**, Consultant, UK, **J.E. Harries**, Imperial College of Science, Technology and Medicine, London, UK, **P.K. Hopke**, Clarkson University, UK, **Sir J. Houghton**, Meteorological Office, UK, **N.J. King**, Environmental Consultant, UK, **P. Leinster**, Environment Agency, UK, **J. Lester**, Imperial College of Science, Technology and Medicine, UK, **S. Matsui**, Kyoto University, Japan, **D.H. Slater**, Consultant, UK, **T.G. Spiro**, Princeton University, USA, **D. Taylor**, AstraZeneca plc, UK, **T.L. Theis**, Clarkson University, USA, **Sir F. Warner**, Consultant, UK.

TITLES IN THE SERIES:

1. Mining and its Environmental Impact
2. Waste Incineration and the Environment
3. Waste Treatment and Disposal
4. Volatile Organic Compounds in the Atmosphere
5. Agricultural Chemicals and the Environment
6. Chlorinated Organic Micropollutants
7. Contaminated Land and its Reclamation
8. Air Quality Management
9. Risk Assessment and Risk Management
10. Air Pollution and Health
11. Environmental Impact of Power Generation
12. Endocrine Disrupting Chemicals
13. Chemistry in the Marine Environment
14. Causes and Environmental Implications of Increased UV-B Radiation
15. Food Safety and Food Quality
16. Assessment and Reclamation of Contaminated Land
17. Global Environmental Change
18. Environmental and Health Impact of Solid Waste Management Activities
19. Sustainability and Environmental Impact of Renewable Energy Sources
20. Transport and the Environment
21. Sustainability in Agriculture

How to obtain future titles on publication

A subscription is available for this series. This will bring delivery of each new volume immediately on publication and also provide you with online access to each title via the Internet. For further information visit www.rsc.org/issues or write to:

Sales and Customer Care, Royal Society of Chemistry
Thomas Graham House, Science Park, Milton Road, Cambridge CB4 0WF, UK
Telephone: +44(0) 1223 432360, Fax: +44(0) 1223 426017, E-mail: sales@rsc.org

ISSUES IN ENVIRONMENTAL SCIENCE AND TECHNOLOGY

EDITORS: R.E. HESTER AND R.M. HARRISON

22
Chemicals in the Environment
Assessing and Managing Risk

RSC Publishing

ISBN 0-85404-206-7
ISSN 1350-7583

A catalogue record for this book is available from the British Library

© The Royal Society of Chemistry 2006

All rights reserved

Apart from fair dealing for the purposes of research for non-commercial purposes or for private study, criticism or review, as permitted under the Copyright, Designs and Patents Act 1988 and the Copyright and Related Rights Regulations 2003, this publication may not be reproduced, stored or transmitted, in any form or by any means, without the prior permission in writing of The Royal Society of Chemistry, or in the case of reproduction in accordance with the terms of licences issued by the Copyright Licensing Agency in the UK, or in accordance with the terms of the licences issued by the appropriate Reproduction Rights Organization outside the UK. Enquiries concerning reproduction outside the terms stated here should be sent to The Royal Society of Chemistry at the address printed on this page.

Published by The Royal Society of Chemistry,
Thomas Graham House, Science Park, Milton Road,
Cambridge CB4 0WF, UK

Registered Charity Number 207890

For further information see our web site at www.rsc.org

Typeset by Macmillan India Ltd, Bangalore, India
Printed by Biddles Ltd, King's Lynn, Noifolk, UK

Preface

Industrial chemicals play a large and often poorly appreciated role in many aspects of life. While members of the general public may realise that many plastics are manufactured by the chemical industry from petroleum-derived feedstocks, they are probably unaware that before such plastics are incorporated into products there may well be a need for addition of stabilisers and/or plasticisers, which are also industrial chemicals. Similarly, while they appreciate that the lubricating oil in their car engine probably derives from the fractionation of crude oil, they are far less likely to appreciate the large numbers of chemical additives which are key to ensuring engine life and a lengthy period between oil changes. It is perhaps because of the widespread ignorance of the general public over the beneficial aspects of chemicals use that they have sometimes been influenced by the environmental pressure groups who tend to focus solely on the problems which can arise through human exposure to such chemicals and through dispersal within the environment. A particular current example is that of flame retardants. Many of the flame retardant chemicals currently in use are of very high bromine content and consequently are highly persistent in the environment. In the absence of obviously superior and environmentally benign alternatives, society is faced with a difficult trade-off between protecting people from the risks of fire and protecting the environment from highly persistent and bio-accumulative chemicals. Policy decisions in such an area are never easy since the benefits and disbenefits are so different in character that is very difficult to determine which outweighs the other, and ultimately for most people this will be a matter of personal opinion.

Given that there are very clear examples of the damaging effects which chemicals can have upon human health and the environment, it is perhaps surprising that for the vast majority of chemicals there is currently little regulation of their manufacture and use beyond that applied to the manufacturing plant under the Pollution Prevention and Control regime. For certain types of chemicals, most notably those used as pesticides, there are positive approval procedures which means that a new pesticide cannot be marketed before it has been formally assessed by an independent body in terms of its human toxicity and environmental effects. However, for the vast majority of chemicals there are currently no such restrictions and the main requirement on the manufacturer is to label the bulk chemical and to provide safety datasheets to the intermediate user. The presumption has been that chemicals can be manufactured, distributed and used unless they are shown to have serious consequences for the manufacturing workforce, the users or the environment. One such example is the use of alkyllead additives in petrol, which started in the 1920s and only ceased in Europe in 2000 as a result of concern over the adverse health effects of population exposures to lead from this and other sources.

Unfortunately, concern for the environment has rarely been seen as an important aspect of product stewardship by the manufacturers of industrial chemicals. For this reason many countries both individually and in collaboration have adopted procedures for assessing the risk posed by industrial chemicals, with a view to implementing either voluntary or compulsory risk management measures where the risks to human health or the environment are considered unacceptably high. Such assessments are typically based upon three criteria, *i.e.* the persistence in the environment, the bio-accumulative tendencies and the toxicity of chemicals to environmental organisms. It is perhaps surprising that for a very large proportion of high production volume chemicals, the basic data from which to assess these key properties is lacking, which has stimulated both new programmes of measurements, but also numerical means of estimating environmental impacts based upon key physico-chemical properties. In very many cases, there are few if any environmental measurements of specific industrial chemicals, and conceptual models have a very important part to play in estimating environmental distributions and concentrations.

While activity in this area has been very high on both sides of the Atlantic, a large stimulus has been provided within Europe by the European Commission's proposal known as REACH (Registration, Evaluation, Authorisation and Restriction of Chemicals), which aims to set a new regulatory regime for both existing substances and newly manufactured chemicals. These proposals have proved controversial, with industry claiming that they will place an undue financial burden on manufacturers, therefore reducing the competitiveness of European industry, and wildlife groups, while welcoming the general thrust of proposals, concerned about the very large amounts of animal testing implied by the need to evaluate toxicity of large numbers of chemicals.

This volume of Issues follows the pattern of many other volumes in the series by including detailed technical articles concerned with certain aspects of a problem along with more discursive opinion-based articles by leading people within the field. The first chapter, by John Garrod of the Department for Environment, Food and Rural Affairs of the UK government, deals with the current regulatory regime for environmental chemicals, summarising the very large volume of regulation currently in place as well as the new proposals from the European Commission. The following chapter by Elliot Finer is more discursive, placing the current procedures and new proposals into a broader context and setting out the benefits and dis-benefits of tighter regulations for the risk assessment and control of chemicals. This is followed by a chapter by Peter Floyd of Risk and Policy Analysts Limited, taking a forward view on how such regulations may develop into the future. It appears that progression of the currently developing methods is largely inevitable.

Having dealt with the legislative, administrative and societal aspects, the volume turns to technical issues and a chapter by Paul Harrison and Philip Holmes of the Institute for Environment and Health describes the means for assessing risk of chemicals to human health, while the following chapter by Lorraine Maltby of the University of Sheffield takes a similar look but in terms of assessing risk to the environment. The last two articles focus more upon the physico-chemical properties of chemicals and their implications. Peter Campbell, Peter Chapman and Beverly Hale deal with risk assessment of metals in the environment. This is a topic requiring particular attention

Preface   vii

to physico-chemical properties since the speciation of metals in environmental media is key to their distribution and biological effects. The final chapter by Don Mackay, Todd Gouin and Eva Webster describes the methods which Professor Mackay and colleagues have pioneered over several decades for assessing the environmental partitioning and persistence of chemicals based upon their physico-chemical properties. Such methods are applicable primarily to organic chemicals and thus this chapter complements that by Campbell and co-workers.

We believe that this volume is timely in view of the current discussion internationally over the extent to which assessment and risk management measures are required for industrial chemicals. Furthermore, we are delighted to have engaged some of the foremost workers in both the technical and policy areas to contribute articles to this volume setting out their view of their technical fields which are key to the future development of the subject.

R.M. Harrison
R.E. Hester

Contents

The Current Regulation of Environmental Chemicals   1
John Garrod
  1 Introduction   1
  2 Protection of the Water Environment   2
  3 Classification and Labelling of Chemicals   6
  4 The Notification of New Substances Regulations   7
  5 Existing Substances Regulations   8
  6 New European Chemicals Policy (REACH)   10
  7 Other European and UK Regulations on Chemicals   11
  8 International Activities on Chemicals   13
  9 Voluntary Approaches to Control of Chemicals in the Environment   15
  References   18

Chemicals Risk Assessment and Management   21
Elliot G. Finer
  1 Introduction   21
  2 Types of Risk   24
  3 Risks and Hazards   26
  4 The Evidence for Harm Caused by Industrial Chemicals   29
  5 Cost–Benefit   34
  6 Perception of Chemical Risks and The Roles of the Advocates   36
  7 The Problems in Controlling Risks from Chemicals   38
  8 Industry Initiatives   39
  9 REACH   40
 10 Conclusions   42
  References   43

Future Perspectives in Risk Assessment of Chemicals   45
Peter Floyd
  1 Introduction   45
  2 Difficulties in Risk Assessment   49
  3 Current Developments   54
  4 Future Perspectives   59
  5 Conclusions   61
  References   61

Assessing Risks to Human Health — 65
Paul T.C. Harrison and Philip Holmes
1 Introduction — 65
2 Chemical Hazard Assessment — 66
3 Assessing Risk — 67
4 Qualitative vs. Quantitative Risk Assessment — 68
5 Role of Epidemiology — 70
6 Use of Biomarkers for Exposure and Risk Assessment — 72
7 Applications of Molecular Biology — 73
8 Uncertainties in Human Health Risk Assessment and the Role of Expert Committees — 75
9 The Changing Face of Chemical Regulation in Europe — 77
10 Future Perspectives — 80
References — 80

Environmental Risk Assessment — 84
Lorraine Maltby
1 What is Environmental Risk Assessment? — 84
2 What are We Trying to Protect? — 89
3 What is the EU Legislative Framework? — 93
4 What are Some of the Challenges Associated with the Environmental Risk Assessment of Chemicals? — 96
References — 99

Risk Assessment of Metals in the Environment — 102
Peter G.C. Campbell, Peter M. Chapman and Beverley A. Hale
1 Introduction — 102
2 Ecological Risk Assessment — 103
3 Comparison of Inorganic and Organic Contaminants — 104
4 Problems with the Application of Traditional ERA Approaches to Metals — 107
5 Conclusions — 123
Acknowledgements — 126
References — 126

Partitioning, Persistence and Long-Range Transport of Chemicals in the Environment — 132
Donald Mackay, Eva Webster and Todd Gouin
1 Introduction — 132
2 Partitioning — 134
3 Persistence — 142

4 Long-Range Transport 148
5 Conclusions 150
Acknowledgements 150
References 150

Subject Index 154

Editors

Ronald E. Hester, BSc, DSc(London), PhD(Cornell), FRSC, CChem

Ronald E. Hester is now Emeritus Professor of Chemistry in the University of York. He was for short periods a research fellow in Cambridge and an assistant professor at Cornell before being appointed to a lectureship in chemistry in York in 1965. He was a full professor in York from 1983 to 2001. His more than 300 publications are mainly in the area of vibrational spectroscopy, latterly focusing on time-resolved studies of photoreaction intermediates and on biomolecular systems in solution. He is active in environmental chemistry and is a founder member and former chairman of the Environment Group of the Royal Society of Chemistry and editor of 'Industry and the Environment in Perspective' (RSC, 1983) and 'Understanding Our Environment' (RSC, 1986). As a member of the Council of the UK Science and Engineering Research Council and several of its subcommittees, panels and boards, he has been heavily involved in national science policy and administration. He was, from 1991 to 1993, a member of the UK Department of the Environment Advisory Committee on Hazardous Substances and from 1995 to 2000 was a member of the Publications and Information Board of the Royal Society of Chemistry.

Roy M. Harrison, BSc, PhD, DSc(Birmingham), FRSC, CChem, FRMetS, Hon MFPH, Hon FFOM

Roy M. Harrison is Queen Elizabeth II Birmingham Centenary Professor of Environmental Health in the University of Birmingham. He was previously Lecturer in Environmental Sciences at the University of Lancaster and Reader and Director of the Institute of Aerosol Science at the University of Essex. His more than 300 publications are mainly in the field of environmental chemistry, although his current work includes studies of human health impacts of atmospheric pollutants as well as research into the chemistry of pollution phenomena. He is a past Chairman of the Environment Group of the Royal Society of Chemistry for whom he has edited 'Pollution: Causes, Effects and Control' (RSC, 1983; Fourth Edition, 2001) and 'Understanding our Environment: An Introduction to Environmental Chemistry and Pollution' (RSC, Third Edition, 1999). He has a close interest in scientific and policy aspects of air pollution, having been Chairman of the Department of Environment Quality of Urban Air Review Group and the DETR Atmospheric Particles Expert Group as well as a member of the Department of Health Committee on the Medical Effects of Air Pollutants. He is currently a member of the DEFRA Air Quality Expert Group, the DEFRA Advisory Committee on Hazardous Substances and the DEFRA Expert Panel on Air Quality Standards.

Contributors

P.G.C. Campbell, INRS Eau, Terre et Environnement, Université du Québec, 490 rue de la Couronne, Québec, QC, Canada G1K 9A9; Tel.: 418-654-2538; e-mail: peter.campbell@ete.inrs.ca.

P.M. Chapman, EVS-Golder, 195 Pemberton Avenue, North Vancouver, BC, Canada V7P 2R4.

E.G. Finer, Southgate, London, UK; Tel.: 020 8886 2941; e-mail: elliotfiner@hotmail.com.

P. Floyd, Risk & Policy Analysts Ltd, Farthing Green House, 1 Beccles Road, London, Norfolk NR14 6LT; Tel.: +44 1508 528465; e-mail: pete@rpaltd.co.uk.

J. Garrod, Strategy and Risk Management, Chemicals and GM Policy Division, Department for Environment, Food and Rural Affairs, 3/E5 Ashdown House, 123 Victoria Street, London SW1E 6DE.

T. Gouin, Canadian Environmental Modelling Centre, Trent University, 1600 West Bank Drive, Peterborough, Ontario, K9J 7B8, Canada; Tel.: 705-748-1005; e-mail: tgouin@trentu.ca.

B.A. Hale, Land Resource Science, University of Guelph, Guelph, ON, Canada N1G 2W1.

P.T.C. Harrison, Director, Institute of Environment and Health, Cranfield University, Silsoe, Bedfordshire, MK45 4DT; e-mail: paul.harrison@cranfield.ac.uk.

P. Holmes, Head of Environmental Toxicology Group, Institute of Environment and Health, Cranfield University, Silsoe, Bedfordshire, MK45 4DT; e-mail: phil.holmes@cranfield.ac.uk.

D. Mackay, Director, Canadian Environmental Modelling Centre, Trent University, 1600 West Bank Drive, Peterborough, Ontario, K9J 7B8, Canada; Tel. 705-748-1005; e-mail: dmackay@trentu.ca.

L. Maltby, Department Animal and Plant Sciences, The University of Sheffield, Western Bank, Sheffield, S10 2TN; Tel.: 0114 222 4827; e-mail: l.maltby@Sheffield.ac.uk.

E. Webster, Assistant Director, Canadian Environmental Modelling Centre, Trent University, 1600 West Bank Drive, Peterborough, Ontario, K9J 7B8, Canada; Tel.: 705-748-1005; e-mail: ewebster@trentu.ca.

The Current Regulation of Environmental Chemicals

JOHN GARROD

1 Introduction

Man-made chemicals and naturally occurring chemicals used in industrial processes or in consumer products may enter the environment at any or all stages of their life cycle from the stage of production/manufacture through formulation and use and including disposal of a used product. The potential for harm to be caused to the environment or to human health via the environment will be dependent on the basic physico-chemical and toxic properties of the chemical, the amount that enters the environment and its distribution between different environmental compartments. The physical properties of a chemical will also determine whether the route of entry is likely to be via air, water or to the terrestrial environment.

The scope of this chapter is largely restricted to industrial chemicals that are not subject to a positive approval system, for example, pesticides, biocides and veterinary medicines, but these may be briefly touched on where appropriate. It will also be restricted to considering the regulation of impacts of chemicals on the environment and on human health via the environment. A consideration of the protection of humans in the work place is also not included.

It is proposed to consider both legal and other frameworks that have been put in place to prevent harm to the environment or to human health via the environment that arises out of the production, use or disposal of hazardous chemicals. Much legislation has been introduced to prevent damage occurring to the environment and human health from chemicals but other approaches including voluntary approaches from government and industry have been adopted.

2 Protection of the Water Environment

Many chemical production facilities have been located historically adjacent to rivers and estuaries frequently to allow ready extraction of water and raw materials (e.g. salt) for industrial processes and for the discharge of waste effluent to the receiving water posing a potential risk to the aquatic environment. The chemical industry has been very successful and thousands of products used in a multitude of commercial and consumer products have been developed and marketed around the world. The UK has become a major player in the international chemical industry. In the past, it was often the case that many chemical production facilities discharged poorly or even untreated waste streams to fresh or marine water bodies resulting in extensive damage to the water quality and resulting in harm to aquatic organisms. In some cases, effluents are discharged directly into the sewage system and pass through a sewage treatment works (STW) before entry into a river or the marine environment. This had the potential to cause harm to the treatment plant itself or for inadequate degradation of the chemical before entering into the receiving water body.

2.1 Control of Chemical Discharges to the Environment

The regulation or control of dangerous chemicals or substances to the environment, particularly to water, from industrial plants or STW has been implemented by various pieces of legislation derived from the Dangerous Substances Directive (76/464/EEC)[1] and Daughter Directives, The Environmental Protection Act (EPA)1990[2] and the Integrated Pollution Prevention and Control (IPPC) Directive (96/61/EEC)[3]. More recently, the Water Framework Directive (WFD) has been introduced to regulate the quality of all water resources in the European Union (EU).

Directive (76/464/EEC) was implemented to control discharges of dangerous substances to inland surface waters, territorial waters, inland coastal waters and ground water. Chemicals identified as the most hazardous to the aquatic environment are in list 1 of the Directive and the legislation requires that these chemicals be completely eliminated from discharges to water. The chemicals in this list are toxic and persist in the environment and accumulate in biological systems causing harm to aquatic life. Dangerous substances in list 2 were thought to be less hazardous than those in list 1. Chemicals in list 2 have an Environmental Quality Standard (EQS) set according to a standard methodology. EQSs represent a concentration limit that must not be exceeded in any controlled water in the UK and the dangerous substance is not believed to be harmful to the aquatic environment at concentrations below this limit.

Where there are uncertainties arising from lack of information, for instance on salt water organisms, larger safety factors are used for the derivation of the EQS. The standards apply to the receiving water and not to the discharge itself. These are statutory standards and the environment agencies (i.e. the Environment Agency, the Scottish Environmental Protection Agency and Environment and Heritage Service, Northern Ireland) have a legal obligation to ensure that they are met. Where special areas for conservation are concerned, the relevant enforcing agency should examine whether an EQS is sufficient to protect the species at the site. The adequacy of EQSs for substances in list 2 has been addressed by Grimwood and Dixon (1997).[4]

Regulatory monitoring for EQS compliance will be carried out at sites from industrial plants and STW where there is the possibility that the discharge effluent will contain dangerous substances. All major rivers will also have background monitoring sites just upstream of their tidal limits, the so-called National Network monitoring sites in the UK.

Discharges to controlled waters from STWs and industrial sites will also be monitored for compliance with their consents to discharge dangerous substances. These consents set limits for individual substance concentrations to ensure that the concentrations in the watercourse downstream do not exceed the EQS.

In 1980 the protection of groundwater was taken out of directive 76/464/EEC and regulated under the separate Council Directive 80/68/EEC[5] on the protection of groundwater against pollution caused by certain dangerous substances.

Broader thinking to address European water policy was addressed in 1988 at the Frankfurt Ministerial seminar on water which resulted in a series of directives after the 1976 Dangerous Substances Directive to fill gaps identified in the existing regulations. This resulted in the introduction of the Urban Wastewater Treatment Directive,[6] the Nitrates Directive in 1991,[7] a Directive for IPPC[8] in 1996 and a new Drinking Water Directive in 1998.[9]

2.2 Water Framework Directive (WFD)

Pressure from the EU Council and the European Parliament to consider water in a more holistic way led to a new European Water Policy, which was developed with extensive consultation with a broad group of stakeholders. The outcome was an agreed way forward for a single piece of legislation to bring together the fragmented pieces of legislation on protecting water. The Commission presented a proposal for a WFD with broad aims. This resulted in a proposal for a single system of management for river basins representing natural features of the hydrological landscape.

Some river basins will cross national boundaries but a 'river basin management plan' will be required and updated every 6 years. With regard to chemical protection, a general requirement for 'good chemical status' was introduced to cover all surface waters. This is defined in terms of compliance with all the quality standards established for chemical substances at European level. The Directive allows for updating existing standards and introducing new ones for priority hazardous substances.

Directive 76/464/EEC will be integrated into the WFD (2000/60/EC).[10] This is a new piece of substantial legislation, which introduces a new integrated approach to the protection of Europe's rivers, lakes, estuaries, coastal waters and groundwater from deterioration and seeks to improve and sustain their quality. The Directive requires all inland and coastal waters to reach 'good status' by 2015. It aims to achieve this through identifying river basins and demanding environmental quality objectives to be set including ecological targets for surface waters.

With regard to dangerous substances, there are transition arrangements from the 1976 Directive. The list 1 of dangerous substances was repealed with the entry in force of the WFD and has been replaced by a List of Priority Substances. Other provisions of the earlier directive, including the reduction of emissions of dangerous substances to controlled waters will continue in place until 2013. The WFD was transposed in the UK through a series of regulations including the Water Environment (WFD) (England and Wales) Regulations 2003[11] and others for specific river basins.

Within the WFD there is also an element of 'good ecological status' which considers the biological, hydrological and chemical characteristics of the catchment area. The biological characteristics will clearly vary from region to region but this is approached by considering what the biology of any particular area would expected to look like in the absence of minimal human impacts although this is not a simple exercise due to the wide range of ecological variability.

In dealing with groundwater, the WFD presumes that the water body should not be polluted at all being a potential source of water for drinking. For that reason, the approach is to prohibit the discharge of any potential pollutant to groundwater and at the same time carry out monitoring to determine ingress of chemicals from indirect sources and take remedial action where necessary.

The new Directive essentially seeks to combine the two aspects of previous legislation that have sought on the one hand to control the entry of hazardous substances into water bodies and on the other to set quality objectives for the receiving environment. It will rationalise the EU's regulations on water by replacing seven existing directives, those on surface water and the related directives on measurement methods, sampling strategies and information requirements on fresh water quality, the fish water, shellfish water and groundwater directives and the directive on dangerous-substance discharges.

2.3 Pollution Prevention and Control Regulations

Pollution from industrial installations is controlled under rules set out in the EC Integrated Pollution Prevention and Control (IPPC) Directive of 1996 (96/61/EC)[8] and this was introduced into England and Wales through the Pollution Prevention and Control(PPC) Regulations 2000[12] and other instruments in Scotland and Northern Ireland. The PPC regulations will replace the pollution control regime set up under Part 1 of the EPA 1990[2] and the transition is due to be completed by 2007.

The EPA introduced to the UK the concepts of Integrated Pollution Control (IPC), which aimed to control releases of hazardous substances to all environmental media, air, land and water and Local Air Pollution Control (LAPC), which controlled releases to air only. The act requires the authorisation to operate the relevant 'prescribed' industrial process from the appropriate regulatory body. Local Councils have responsibility for authorising prescribed processes that emit pollutants to air only (Part B processes) and the Environment Agency or the Scottish Environment Protection Agency authorise processes that may emit pollutants to air, land or water.

In April 2000, Part IIA of the EPA came into force and this introduced a new regime for the regulation of contaminated land. The main purpose is to make provision for the identification of land that poses unacceptable risks to the environment or human health from polluting chemicals and to enforce remediation where the risks cannot be reduced by other means.

The PPC Regulations 2000 introduce three linked systems for pollution control. These relate to the potential of the industrial site or installation to pollute the environment and are graded A (1) (highest) through A (2) to Part B (lowest). A1 sites are regulated by the Environment Agency and A2 and Part B installations are regulated by local authorities.

A (1) installations are subject to IPPC, A (2) installations to Local Authority IPPC (LA-IPPC) and B processes to LA- PPC. All three systems require the operators of the installations to obtain a permit to operate.

IPPC and LA-IPPC introduce an integrated environmental approach to the control of certain industrial processes. The IPPC Directive seeks to achieve a high level of environmental protection and to prevent or reduce to an acceptable level emissions to the environment. A guiding principle under the PPC regulations is the use of 'Best Available Techniques' (BAT), which is designed to balance the cost of compliance to the operator against the benefits to the environment. The PPC regulations increase the scope of control above the EPA regime to cover energy efficiency, site restoration, accident prevention, noise, odour, waste minimisation and heat and vibrations and they cover a wider range of activities to include food and drink manufacturers, intensive live-stock production and landfill sites. LA regulated Part B installations extend only to emissions to air but BAT also applies.

Guidance documents have been published on the PPC regulations to help installation operators to understand and meet the requirements.

2.4 Direct Toxicity Assessment

The control of point source discharges by the regulatory agencies has been predominantly through the consents procedure for individual substances. This procedure is set out in Schedule 10 of the Water Resources Act 1991.[13] A consent may be issued subject to certain conditions dependent on the location of the discharge, the design and construction of the outlet, the composition and quantity of the effluent and sampling requirements.

However, there are some industrial discharges which contain a mixture of components creating a complex effluent that is entering the receiving water. These circumstances place difficulties on setting an appropriate consent for the whole effluent. The difficulties may arise through the potential high cost of analysis of all of the chemical components, scarce data for ecotoxicological end points for many substances and the difficulty in predicting how chemicals will behave in complex combinations. Such a situation calls for an alternative approach to assess the toxicity of the whole effluent and this approach is termed Direct Toxicty Assessment (DTA). To improve the assessment and control of complex effluents, the UK through the Environment Agency and the Scottish Environmental Protection Agency has been developing an approach for DTA.

The DTA approach will use a suite of well-established toxicity tests on algae, invertebrates and, where necessary, fish which can be used as a direct assessment of the biological quality of an effluent or environmental compartment such as the water column or sediment. The testing methods will also permit the identification of toxic waste streams discharging to a watercourse or to sewer and their source within a production facility. In some cases toxicity tests are already used for licensing of discharges and DTA has important implications for regulation as a component of IPC. The Scottish Environmental Protection Agency has issued a Technical Guidance Manual[14] for DTA for licensing discharges to water.

The assessment and control of effluent toxicity currently relates to acute toxicity only. It is acknowledged that the approach could be extended to the assessment of chronic toxicity and endocrine disrupting effects. The combined use of biological effects measures, which includes the use of biomarkers, substance-specific assessment and biological surveillance will provide a robust system for identifying, characterising and controlling the effects of hazardous chemicals on the environment.

3 Classification and Labelling of Chemicals

In the 1960s, the European Economic Community(EEC) recognised a need to provide protection from dangerous substances of human health and in particular workers handling them. This resulted in the publication in 1967 of Directive 67/548/EEC (The Dangerous Substances Directive)[15] relating to the classification, packaging and labelling of dangerous substances and provides another mechanism for managing the risk to the environment and human health from chemicals. The Directive covers all substances and preparations which are placed on the market in Member States of the Community and substances are defined as chemical elements and their compounds either in their natural state or as produced by industry. The original Directive did not require labelling substances as 'dangerous for the environment' but this was introduced with the sixth amendment of the Directive in 1979.

In the UK, the Chemicals (Hazard Information and Packaging for Supply) Regulations (CHIP)[16] implement the Dangerous Substances Directive and later directives including 1999/45/EC[17] relating to dangerous preparations and the Safety Data Sheets Directive (91/155/EEC and related directives).[18] Under CHIP, substances have always required an environmental classification but the latest set of regulations, CHIP 3, which came into force in July 2002, requires both substances and preparations to be labelled if they have the potential to cause harm to the environment. Preparations are defined as mixtures or solutions composed of two or more substances so there may be considerable implications for the new requirement under CHIP 3. The CHIP regulations require those who supply dangerous chemicals to identify the hazards of the chemical, to provide this information to downstream users and customers via the label or in a safety data sheet and to package the chemical safely. In the context of CHIP, 'supply' is used in a broad sense wherever a chemical is supplied to another person and can include retailers, wholesalers, distributors, importers or manufacturers of chemicals.

Guidance on the classification of substances and preparations is provided in Annex VI of the Dangerous Substances Directive which is updated to reflect technical progress. Changes are agreed by experts from Member States known as the Technical Progress Committee (TPC) supported by the European Chemicals Bureau(ECB), which also engages interested stakeholders, including industry, in the discussions. Resulting measures are known as Adaptations to Technical Progress which are formally adopted as a Commission Directive. The TPC is also responsible for Annex I of the Dangerous Substances Directive, which is the published list of substances with a harmonised classification and labelling which contains information on approximately 8000 substances.

The classification currently has 15 categories of danger including explosive, very toxic, harmful, irritant, carcinogenic and dangerous for the environment. The criteria for the latter are set out in Annex VI of Directive 67/548/EEC[15] and comprise risk phrases indicating toxicity, harm or long-term adverse effects to the aquatic environment.

The regulations are administered in each Member State by the 'Competent Authority'(CA), which in the UK comprises the Health and Safety Executive and the Environment Agency acting jointly.

A Globally Harmonised System(GHS) of Classification and Labelling of Chemicals[19] was adopted in December 2002 and is being implemented worldwide with a target for completion of 2008. The purpose of the GHS is to provide a common system to cover the wide range of languages, ages and social conditions that people face around the world. The intention is to communicate information on the hazards posed by chemicals particularly through the provision of safety data sheets. Agenda 21 adopted at the United Nations Conference on Environment and Development (UNCED) in 1992 provided the mandate to complete this system. It was managed under the auspices of the Interorganisation Programme for the Sound Management of Chemicals (IOMC).

4 The Notification of New Substances Regulations

Notifications schemes for new chemical substances which are manufactured or imported into the EU were first introduced during the 1970s by individual member states. This allowed a system for potential assessment of risks to consumer and occupational health and the environment. A sixth amendment to the Dangerous Substances Directive (DSD) (79/831/EEC)[20] introduced a European notification system in 1981. A seventh amendment to the DSD (92/32/EEC)[21] was adopted in April 1992 with effect from November 1993 and introduced a risk assessment for new notified substances. This Directive was implemented in the UK as the Notification of New Substances Regulations 1993 (NONS 93).[22] Over 6000 notifications have been made since 1981 covering more than 3700 substances of which some 2500 have been notified under the seventh amendment since 1993. The seventh amendment is a single market directive, which ensures that notification requirements are the same in all EC Member States and that a notification accepted in one Member State is valid for all of them. The notification requirements are harmonised across the Member States, which is intended to save notifiers, mostly industrial organisations and companies, time and money.

A notification for a new chemical substance requires the presentation of a technical dossier containing information about the production and uses of the chemical and data about its physical, chemical, toxicological and ecotoxicological properties. There is a requirement for classification and labelling proposals with any necessary precautions that need to be taken and a draft assessment of any risks that the chemicals pose to human health or the environment. The amount of information required increases according to the quantity of chemical manufactured per year starting at 10 kg and with a 'baseset' of data for substances from 1 to 10 tonnes per year and with further toxicological and ecotoxicological testing required for quantities above 10, 100 and 1000 tonnes.

The regulations are administered in each Member State by a Competent Authority (CA), which in the UK is the Health and Safety Executive and the Environment Agency acting jointly. The function of the CA is to evaluate the technical dossiers, provide a risk assessment, request further information from the notifier if necessary and liaise with European Chemicals Bureau (ECB) Work Area New Substances which manages the New Chemicals Data Base. There is a system whereby information exchange and discussion on dossier content and interpretation takes place between Member States through meetings chaired by the ECB or Directorate General (DG) Environment. The outcome of these meetings is recorded in a 'Manual of Decisions,' which provides information and guidance for industrial and other stakeholders with an interest in new substances and their risks to the environment and human health.

The definition of a New Substance is limited by the enforcement date of the sixth amendment to the Dangerous Substances Directive mentioned earlier (79/831/EEC), which differentiates between new and existing substances. Existing substances are listed in the European Inventory of Existing Commercial Chemical Substances (EINECS), which was published in the Official Journal of the European Union on 15 June 1990. The Inventory lists 100,196 entries, which were on the EU market between 1 January 1971 and 18 September 1981. Chemicals introduced after 1981 are listed in the European List of Notified Chemical Substances (ELINCS), which is periodically updated in an Official Journal (OJ). New chemicals in this category do not include pesticides, veterinary medicines, biocides, pharmaceuticals, cosmetics, foodstuffs, radioactive materials, chemical wastes and substances used only in scientific research in small quantities.

5 Existing Substances Regulations

Council Regulation 793/93/EEC[23] was adopted in 1993 and legislates on the evaluation and control of the risks from all existing substances, listed in EINECS, manufactured or imported into the EC. In the UK the Regulation is enforced by the Notification of Existing Substances (Enforcement) Regulations (1994).[24]

Regulation 793/93/EEC was one outcome following concern over potential risks arising from existing chemicals towards the end of the 1980s and the Fourth Community Action Plan for the Environment (1987–1992) of the Council of the European Community highlighted as a major objective the evaluation of the risks to the environment and human health from chemical substances. The Fourth Action Programme proposed a means of prioritising chemicals for action and suggested how appropriate data should be gathered and assessed for risks to the environment and human health.

Just before the EC action programme was launched, the Organisation for Economic Co-operation and Development (OECD) had recognised the need for work on existing chemicals by recommending in 1987 a Systematic Investigation of Existing Chemicals and launched a programme of work in 1988. OECD and United Nations initiatives on chemicals are furthermore linked into the international arena through the development of strategy for the environmentally sound management(ESM) of toxic chemicals which was outlined in Chapter 19 of Agenda 21 of the Rio Declaration and adopted at the United Nations Conference on Environment & Development (UNCED) in 1992.

The Existing Substances Regulation complements the new substances element of the Dangerous Substances Directive Regulation and has four discrete phases to the assessment process; collection of data, setting priorities, risk assessment and risk reduction.

As a means of setting some priorities for dealing with the large number of chemicals in the EINECS, the Regulation initially looked at chemicals produced in high-production volumes (HPVs) as a surrogate for potential environmental exposure. In the context of the regulations, an HPV is a chemical which was either produced or imported into the EC in quantities exceeding 1000 tonnes per year in the period between March 23, 1990 and March 23, 1994. Subsequently, there was a requirement for submission of a reduced data package for chemicals in the 10–1000 tonne bracket by June 4, 1998. The Regulation requires that all data are in a standard format, the Harmonised Electronic Data Set (HEDSET) and all the information is stored in the International Uniform Chemical Information Database (IUCLID), which is required to be updated every 3 years by companies that have previously submitted data which is held on it.

The data collected is reviewed by the Commission and Member States and lists of chemicals requiring priority action drawn up. Four such lists have been drawn up since 1994 comprising 144 chemicals. Prioritised substances undergo an extensive assessment of their hazards and risks to man and the environment, the latter covering, aquatic, terrestrial and the atmospheric compartments and any tendency for chemicals to bioaccumulate in the food chain. The risk assessments are carried out according to the Technical Guidance Documents (TGD) on Risk Assessment for New and Existing Substances.[25] On the basis of the TGD, the ECB has produced a European Union System for the Evaluation of Substances (EUSES),[26] which enables users to assess risks rapidly and efficiently by looking at the relationship between a substance's expected emissions and its effects on organisms.

The outcome of a risk assessment may be categorised in one of three ways:

(i) No need for further information/testing
(ii) At present no need for further information and/or testing and no need for risk reduction measures
(iii) Need for limiting the risks.

Where the outcome is the need for limiting the risks, a risk reduction strategy must be developed. This is carried out according to the Technical Guidance Document (TGD) on Risk Reduction. The implementation of risk reduction strategies is primarily via the Marketing and Use (M&U) Directive (76/769/EEC),[27] which will specify in an amendment to the Directive the restrictions that need to be placed on using the substance in order to reduce the risk and prevent harm. For example, Directive 2003/53/EEC is a recent amendment to the M&U Directive that places restrictions on the use of nonylphenol and nonylphenol ethoxylates where those uses result in discharges, emissions or losses to the environment. The risk assessment, which identified a need to reduce risks, was identified by the Commission in March 2001 and the decision endorsed by the Scientific Committee on Toxicity, Ecotoxicity and the Environment (CSTEE), which has more recently become the Scientific Committee on Health and Environmental Risks (SCHER). Other chemicals recently controlled under the M&U Directive include the brominated flame retardants pentabromodiphenylether and octabromodiphenylether. In the past, substances such as polychlorinated biphenyls (PCBs), asbestos and cadmium have been subject to restrictions on their use via this directive.

6 New European Chemicals Policy (REACH)

It has been widely accepted across the EU Member States that since the Existing Substances Regulations were introduced in 1993, the number of chemicals which have progressed to some form of control or regulation has been slow even though around 140 have been prioritised as of high concern for the environment. This is only a very small proportion of those 100,000 or so chemicals on EINECS of which some 30,000 are thought to be produced or imported into the EU in quantities of over 1 tonne.

The need to reform the Community's chemicals policy was discussed at an informal meeting of EU Environment Ministers in Chester in 1998 and this prompted the Commission to issue a White Paper in February 2001 on a Strategy for a Future EU Chemicals Policy. Following much discussion and dialogue with all major stakeholders, the Commission issued a proposal for a New Regulatory Framework for the Registration, Evaluation and Authorisation of Chemicals (REACH) on 29 October 2003. REACH aims to protect the environment and human health via the environment while at the same time maintaining the competitiveness and enhancing innovation within the chemicals industry in the EU.

REACH is proposed to cover both new and existing substances and will require industrial organisations that manufacture or import chemicals in excess of 1 tonne per annum to register the chemical in a central database.

Information to be provided on the chemical properties will be dependant on the quantity of chemical supplied or imported in a rather similar but less extensive manner to that required under the current UK New Substances Regulations. Emphasis for the registration procedure will be placed on the maximum sharing of data with the aspiration of one registration for each substance. This will have the effect of minimising costs to industry and also will be one way for achieving a key objective for the regulation of minimising the use of animals for the testing that will be necessary to provide the data required to ensure adequate protection of human health and the environment.

The Evaluation stage of REACH will examine the data provided during registration to determine whether any further information is required and the hazards and risks posed by the chemical substance.

Chemicals evaluated as of most concern which are likely to be, at least in the first instance, those produced in quantities of over 1000 tonnes and with certain toxic, persistent and bio-accumulative properties (PBT) or category I and II carcinogens, mutagens or reprotoxins will be subject to Authorisation. This will require the registrant to gain a specific approval for all uses to which any chemical is put. All other uses of the chemical would be banned. Implicit in its approach to dealing with chemicals of high concern, REACH includes the concept of chemical substitution. Substitution will require that, wherever possible, products or processes using chemicals of high concern should replace them with an alternative chemical or process which poses a lower risk to the environment.

The REACH proposals are currently being considered by the Council of the EU and the European Parliament for adoption under the co-decision procedure. The Commission has defined an Interim Strategy leading up to and after the entry into force of the Regulation. The strategy comprises three phases; an interim period running from 2004 to 2006 will put in place a number of preparatory actions to enable the effective administration of the legislation immediately after it comes into force; a transition period covering 2006 –2008 and a final phase commencing in 2008, when the proposed European Chemicals Agency takes over running REACH from the Commission. The interim strategy will prepare for REACH by developing Technical Guidance Documents (TGDs), software tools and infrastructure through a series of REACH Implementation Projects (RIPs), by re-evaluating current work on chemicals to make resources available for the development of REACH and by ensuring full stakeholder engagement, particularly with industry, to ensure that the various components of REACH are workable. Much of this work is the responsibility of the ECB, which provides support for the Commission on this new legislation on chemicals.

In the UK, a Government position statement on REACH was published in December 2002 and since then the Government has, among other activities, responded to the Commission consultation of May 2003, launched its own consultation process in March 2004, held a stakeholder conference in April 2004, carried out a series of regulatory impact assessments and been particularly active in developing a proposal for 'One Substance, One Registration' to optimise data sharing to minimise costs and the requirement for animal testing.

7 Other European and UK Regulations on Chemicals

7.1 OSPAR

The Convention for the Protection of the Marine Environment of the North-East Atlantic of 1992 (the OSPAR Convention),[28] of which the UK were one of 16 contracting parties, agreed that all should take measures to safeguard the marine environment against adverse effects of human activities, including pollution from hazardous chemicals to protect human health and to conserve the marine ecosystem. It entered into force in March 1998. It replaces the Oslo (1972) and Paris (1974) Conventions but decisions, recommendations and all other agreements adopted under those Conventions continue to be applicable unless terminated by new measures adopted under the 1992 Convention.

An OSPAR Commission was established to administer the Convention and to develop policy and international agreements. The Commission has reached a number of decisions, recommendations and other agreements relating to its Hazardous Substances Strategy[29] which was reaffirmed in 2003. The strategy aims to reduce pollution of the marine environment through a continuous programme of reducing discharges, emissions and losses of hazardous substances with the intention of eliminating them by the year 2020.

A List of Substances of Possible Concern to the marine environment has been drawn up and priorities for individual chemicals for risk assessment and any necessary action have been determined. The list is a dynamic list and is regularly revised, most recently in August 2004. The list was drawn up according to a Dynamic Selection and Prioritisation Mechanism for Hazardous Substances (DYNAMEC) developed by the Convention. The list, originally agreed in 2002, contains approximately 400 chemicals that meet specific criteria for persistence, toxicity and bioaccumulation potential. Industries are invited to bring forward data for each prioritised chemical and each is adopted by a sponsor country which, in conjunction with OSPAR, produces a background document covering the properties of the chemical and likely pathways to the marine environment. The document provides an environmental risk assessment and is used to recommend any actions required to meet the aims of the Convention on hazardous substances.

OSPAR works with the European Commission to ensure that work is not duplicated on specific chemicals under other regulatory programmes (such as the Existing Substances Regulations) and carries out hazard and risk assessments according to the same guidelines as those used under the Existing and New Substances Regulations described earlier.

7.2 The UK Offshore Chemicals Notification Scheme and the Offshore Chemical Regulations

OSPAR also played a key role in the development of the Offshore Chemicals Notification Scheme(OCNS), which was originally introduced in 1979, and applies to all chemicals that are used in connection with the exploration and other offshore activities in the processing of petroleum and gas on the UK Continental Shelf. The UK Government introduced a revised scheme in 1993 to take account of test protocols approved by OSPAR and this was taken forward further in 1996 when OSPAR introduced a Harmonised Offshore Chemical Notification Format (HOCNF), which is used to define the testing requirements for offshore chemicals in the NE Atlantic sector.

OSPAR introduced a decision in 2002 on a Harmonised Mandatory Control System for the Use and Reduction of the Discharge of Offshore Chemicals, which are administered in the UK under the Offshore Chemical Regulations (OCR) 2002[30] which came into force on 15 May 2002. The Department for Trade and Industry administers the OCR with technical support from Defra's Centre for Environment, Fisheries and Aquaculture Science (CEFAS).

The regulations have been introduced to ensure better control, in a proportionate and cost-effective manner, of the use and discharge of offshore chemicals which inevitably contaminate the marine environment. The substitution of hazardous for lower impact chemicals is also a key element of these regulations.

The system for risk assessment of chemicals is similar to that used in other chemical regulatory regimes but uses data on marine organisms (algae, invertebrates and fish) and chemicals are ranked on the basis of a Hazard Quotient and their persistence and bioaccumulative properties. The regulation also places an obligation on authorities to use the Chemical Hazard Assessment and Risk Management (CHARM) hazard assessment module as the primary tool for ranking chemicals. Chemicals are placed on a List of Notified Chemicals for 3 years when they become due for re-certification. The hazard data are used by operators of exploration platforms to carry out site-specific risk assessments for the particular chemicals they propose to use.

7.3 Hazardous Waste Regulations

Hazardous chemicals can become hazardous waste at the end of their life and the issue of how to dispose of, transport, store or treat those wastes is addressed by various regulations and directives within the EU. The European Hazardous Waste Directive (91/689/EEC)[31] sets out the requirements for the controlled management of hazardous (special) waste. The Directive was implemented in England, Scotland and Wales by the Special Waste Regulations (1996)[32] and various subsequent amendments to them. The regulations were substantially reviewed in 2000[33] and improvements have been developed and, following a consultation in 2004 on new Hazardous Waste Regulations for England, it is intended to introduce them during 2005. One particular requirement of the new regime will be for producers of hazardous waste to notify their premises to the Environment Agency and guidance on this process will be provided.[34]

8 International Activities on Chemicals

8.1 OECD and ICCA HPV Chemicals Programmes

In 1990, OECD decided to embark on a system for evaluating the hazards from High Production Volume (HPV) Chemicals that would involve the co-operation and sponsorship of all member countries. HPV chemicals are defined as those chemicals reported to be produced or imported into member countries of OECD or the EU in quantities of over 1000 tonnes per year. Industry would be encouraged to provide the information needed to complete a Screening Information Data Set (SIDS) with an agreed number of initial toxicological and ecotoxicological end points. When the SIDS dossier is complete, an initial appraisal of the data is made (SIAR, SIDS Initial Assessment Report), potential hazardous properties of the chemical are identified and areas for further work are recommended if need be. Exposure information is taken into consideration based largely on the use to which the chemical is put to make a preliminary assessment of likely risks to the environment and human health. SIDS dossiers and assessments are made available worldwide through the United Nations Environment Programme (UNEP) when they have been agreed by all member countries at a SIDS Initial Assessment Meeting (SIAM).

In 1998 a major refocusing of the HPV programme took place to streamline various aspects of the selection of chemicals, to enhance the SIDS testing programme and to focus the SIDS work on hazard identification but to leave risk assessment to be looked at by a joint OECD/IPCS (International Programme on Chemical Safety) project.

Also in 1998 a major voluntary initiative was launched by the International Council of Chemical Associations (ICCA) to provide internationally harmonised data sets and hazard assessments for approximately 1000 HPV Chemicals. This programme effectively replaced the system by which member states sponsored individual chemicals through the system. The ICCA has been working in partnership with the refocused OECD HPV programme and using the OECD HPV Chemicals List to establish priority chemicals for action. The aim was to have SIDS completed by the end of 2004 but while this has not been met with only around 260 being assessed to date, all 1000 are being progressed. The ICCA are urging participating companies to complete work on the remaining chemicals as soon as practicable. Chemical companies have been providing data needed for the various physical and chemical properties and toxicological and ecotoxicological end points required under the OECD programme. Industry recognises the value of the availability of the hazard data to internationally agreed guidelines to prevent duplication of testing, minimising the numbers of animals used for testing and costs to themselves and that it provides a framework for the risk assessment of the chemicals on a global, regional or national basis.

8.2 United Nations Environment Programme (UNEP)

In addition to work carried out on chemicals by the OECD, the United Nations co-ordinates a wide range of international activities to control hazardous chemicals. The Stockholm Convention of 2001[35] is a worldwide treaty to protect the environment and human health from persistent organic pollutants (POPs). These chemicals, which include organochlorine insecticides such as DDT, aldrin and dieldrin, toxaphene, PCBs and dioxins and furans, are chemicals which persist in the environment for long periods of time, have the properties to become very widely distributed around the globe, bioaccumulate in the bodies of living organisms and are toxic to wildlife and humans. An assessment report on these chemicals was produced for the International Programme on Chemical Safety in 1996 and served as a basis for development of a work plan to complete the assessment process called for in the UNEP.

The Stockholm Convention, an international legally binding instrument, implemented international action on POPs and national governments are taking measures to reduce the release of POPs into the environment. The Convention entered into force in May 2004 and was ratified by EU in November 2004 and by the UK in January 2005. Regulation 850/2004/EEC[36] bans the intentional production, marketing and use of the substances listed in the Convention so far. The first Conference of the Parties of the Stockholm Convention was held in May 2005 to ensure effective implementation and further development of the Convention. One aim will be to set up a POPs Review Committee for adding new chemicals to the Convention. The adoption of guidelines on the minimisation of dioxin releases and the consideration of guidelines on the levels of POPS permitted in waste will also be considered.

In 1998 governments adopted the Rotterdam Convention[37] of the United Nations, which made the Prior Informed Consent (PIC) Procedure for Certain Hazardous Chemicals and Pesticides in International Trade legally binding. From 1980 up to 1998, the PIC procedure was voluntary and required exporters of hazardous chemicals to obtain the PIC of importers before proceeding with the trade. The Rotterdam Convention gave importing countries the means to identify potentially hazardous chemicals and to withhold importation of any that they were unable to manage safely. Where hazardous chemicals are imported, the convention provides for support in terms of appropriate labelling and handling. It also enforces compliance on exporters to meet the requirements of PIC. The Rotterdam Convention entered into force on 24 February 2004 after the EU had implemented in January 2003 Regulation 304/2003/EEC[38] concerning the export and import of dangerous chemicals. The first conference of the parties to the Convention held in September 2004 agreed the addition of 14 hazardous chemicals to Annex III of the Convention and established a chemicals review committee.

A decision to create a Strategic Approach to International Chemicals Management (SAICM) was adopted by the UNEP Governing Council in February 2002. UNEP agreed to work with member governments and other stakeholders to review current actions to advance the sound international management of chemicals, to further advance the approach of the Intergovernmental Forum on Chemical Safety (IFCS) Bahia Declaration on Chemical Safety and to propose specific projects and give them priorities. It is due to be adopted at an International Conference on Chemicals Management (ICCM) in February 2006.

The Bahia Declaration of 2000[39] set out some key goals for chemicals in the international arena. These were centred on the essential role of robust chemical management in sustainable development and the protection of human health and the environment and promoted a precautionary approach. Other objectives were to assist countries with issues of chemical safety, better provision of data on chemicals and control of illegal trade in hazardous chemicals.

8.3 Disposal of Hazardous Waste

The Basel Convention[40] adopted on 22 March 1989 in response to international concern about the shipping of hazardous waste, including waste chemicals, to developing countries when a tightening of legislation in the 1980s resulted in a steep rise in the cost of disposal. The Convention was initially concerned with setting up a mechanism to control the movement of hazardous waste across national boundaries and developed criteria for Environmentally Safe Management (ESM) of hazardous wastes. The Convention is now building on the framework for movement control to implement and enforce treaty commitments. It will also focus on the minimisation of hazardous waste generation.

The objective of ESM is to protect human health and the environment by minimising hazardous waste production. The concept invokes an 'integrated life cycle' approach where controls are placed at all stages in waste generation through to transport, treatment, re-use, recycling, recovery and final disposal.

A guiding principle of the Basel Convention is that hazardous wastes should be dealt with close to where they are produced. Under the Convention movement of hazardous waste across international borders can take place only upon prior written agreement between the state of export and the states of import. It is illegal to make shipments without such agreement and without appropriate documentation. An exception is granted where an exporting state does not have the capacity to deal with the waste in an environmentally acceptable manner.

The Secretariat of the Convention co-operates with individual national authorities to help them in developing national legislation, compiling waste inventories, assessing hazards and preparing waste management plans and policy tools. It also provides advice on dealing with accidents with hazardous waste and can provide expertise and equipment if necessary to national authorities.

9 Voluntary Approaches to Control of Chemicals in the Environment

9.1 UK Chemicals Strategy

In 1999 the UK Government set out its proposals for a Chemical Strategy on the Sustainable Production and Use of Chemicals.[41] Its aim was to avoid harm to the environment and human health via environmental exposure to chemicals. The strategy was designed to make information about the environmental risks of chemicals publicly available, to continue to reduce the risks to the environment from chemicals while maintaining the competitiveness of industry and to phase out those chemicals that pose an unacceptable risk to the environment and human health. A key feature of the strategy was the establishment of the UK Chemicals Stakeholder Forum (CSF) to promote a better understanding between stakeholders on issues of chemicals and the environment and to provide advice to the Government about chemicals in the environment to guide the development of policy. The Stakeholder Forum was established in 2000 with the following terms of reference:

- Act as a barometer for the views and opinions of stakeholders
- Advise on the selection of criteria for identification of chemicals needing priority attention and the risk management strategies required for them
- Report on such chemicals, advising on precautionary controls and restrictions and time scales for action
- Advise on the development of indicators of environmental exposure to hazardous chemicals, including targets for reducing overall exposure of the environment
- Conduct its business in an open fashion, making its documents, minutes and advice public.

The work of the CSF has been carried out in conjunction with technical support from the Environment Agency and the Advisory Committee on Hazardous Substances (ACHS), which added support to the CSF to its terms of reference in 2000. The ACHS is an independent Scientific Non-Departmental Public Body made up of 10 experts in the fields of toxicity, ecotoxicity and the properties and behaviour of chemicals in the environment.

An early task of the Stakeholder Forum was to decide on a set of criteria for chemical properties to enable selection of substances for priority action. The criteria were based on properties of likely persistence (P) in the environment, the potential to accumulate in biological systems (B) and toxicity (T) to environmental organisms or to humans. Chemicals meeting the criteria were then identified from HPV chemicals in the International Uniform Chemical Information Database (IUCLID) that was set up under the Existing Substances Regulations referred to earlier. This resulted in a list of approximately 120 chemicals potentially having PBT properties and the CSF 'list of chemicals of concern' was published on the internet in June 2003.

The CSF has selected chemicals from this list and engaged with industry to progress actions to gather data on hazardous properties and, where appropriate, asked for action to be taken to reduce risks to the environment. An early success of the CSF was to recommend a voluntary agreement between industry and the Government for the reduction of risks from nonylphenol, nonylphenol ethoxylates and to prevent any increase in risks from octylphenol and octylphenol ethoxylates and this was agreed in April 2004 with the chemical supply industry and downstream users.

Between 2002 and 2004, the CSF considered 10 chemicals on its list in some detail including tetraethyl lead, hexamethyldisiloxane, vinylneodecanoate, hexabromocyclododecane, tetrabromobisphenol-A, dodecylphenol and tertiary-dodecylmercaptan.

These chemicals are now all part of ongoing programmes of hazard and risk assessment and, where appropriate, risk reduction and progress is regularly reviewed by the Stakeholder Forum at its 3-monthly meetings. All of the work of the CSF is published in an Annual Report.

During 2004, the work of the CSF had a major review, which included an external consultation, and Ministers subsequently agreed a change of focus for the CSF work to include an advisory role to Government on the developing REACH regulation. The review also led to a reconsideration of how to speed up the consideration of individual chemicals and this resulted in moving from single substance consideration to looking at groups of chemicals at the same time. The ACHS identified nine groups from the CSF list of chemicals of concern and three of those, representing around 20 chemicals are now being considered by the CSF. The prime objective is to establish the PBT properties of the chemicals to determine whether they are of high concern and should be candidates for further action. Industry, in the form of individual companies and chemical sector organisations assist the CSF in the provision of data which feed in to European or international chemical programmes on hazard and risk assessment as described earlier. The UK Environment Agency does undertake National Risk Assessments, according to internationally agreed guidelines, to speed up action on particularly hazardous chemicals that have been identified as of concern to the UK environment and human health. Some of the chemicals from the CSF list of concern are being evaluated through this route which can lead to recommendations for local risk reduction measures being taken where unacceptable risks have been identified.

Since October 2004, the CSF has addressed two major issues that are part of the developing REACH regulation. It has considered the UK Government position papers on 'One Substance One Registration' and 'Substitution', the latter a crucial aspect of the process of chemical authorisation proposed under REACH. The CSF will continue to address further issues as the policy develops and the regulation moves towards the implementation phase.

9.2 *Royal Commission on Environmental Pollution*

The Royal Commission on Environmental Pollution is an independent standing body which was established in 1970 to advise the Queen, the Government and members of the public on environmental issues. Its 24th report published in June 2003 was entitled 'Chemicals in Products; Safeguarding the Environment and Human Health'[42] and it sought to challenge current approaches to the regulation of chemicals and to make recommendations for future regulatory systems. The 54 recommendations focussed on a new approach to the management of chemicals in the environment some of which may be common to or complementary to the REACH proposals. The recommendations comprise a stepwise system for handling chemicals commencing with the compilation of a list of all marketed chemicals, sorting, selecting and evaluating chemicals of concern and introducing risk management action where necessary. The report emphasises the need to phase out animal experiments in regulatory testing and promotes the greater use of *in vitro* testing methods and the wider application of existing data and computational techniques to describe the hazardous properties of chemicals.

The Commission also considered aspects of how to manage a chemical assessment programme, international harmonisation of processes and procedures and various ways to encourage action from industry through, for example, the introduction of a charging scheme to encourage chemical substitution, further voluntary initiatives like those promoted by the CSF, better information supply and labelling and improvements to the regulatory process to stimulate innovation in the chemical industry.

References

1. 76/464/EEC, Council Directive of 4 May 1976 on pollution caused by certain dangerous substances discharged into the aquatic environment of the community. Official Journal of the European Union (OJ), 1976, **L129**, 23–29.
2. Environmental Protection Act, HMSO, London SW8 5DT, 1990, ISBN 0-10-544390-5.
3. 96/61/EEC, Council Directive of 24 September 1996 concerning integrated pollution prevention and control. OJ, 1996, **L257**, 26–40.
4. M.J. Grimwood and E. Dixon, Assessment of risks posed by List II metals to sensitive marine areas (SMAs) and adequacy of existing environmental quality standards for SMA protection. Report to English Nature, 1997.
5. 80/68/EEC, Council Directive of 17 December 1979 on the protection of groundwater caused by certain dangerous substances. OJ, 1980, **L20**, 43–48.
6. 91/271/EEC, Council Directive of 21 May 1991 concerning urban wastewater treatment. OJ, 1991, **L135**, 40–52.
7. 91/676/EEC, Council Directive of 12 December 1991 concerning the protection of water against pollution caused by nitrates from agricultural sources. OJ, 1991, **L375**, 1–8.
8. 96/61/EEC, Council Directive of 24 September 1996 concerning integrated pollution prevention and control. OJ, 1996, **L257**, 26–40.
9. 98/83/EEC, Council Directive of 3 November 1998 on the quality of water for human consumption. OJ, 1998, **L330**, 32–54.
10. 2000/60/EEC, European Parliament and Council Directive of 23 October 2000 establishing a framework for the Community action in the field of water policy. OJ, 2000, **L329**, 1–73.
11. The Water Environment (Water Framework Directive)(England and Wales) Regulations, Statutory Instrument 2003, No. 3242, HMSO, London, 2003, ISBN 0-11-048355-3.
12. The Pollution Prevention and Control (England and Wales)(Amendment) Regulations, HMSO, London, 2002, ISBN 0-11-039380-5.
13. Water Resources Act, HMSO, London, 1991, (c57), ISBN 0-10-545791-4.
14. Direct Toxicity Assessment, Technical Guidance Manual for Licensing Discharges to Water, Guidance No. 03-DLM-COPA-EA3, Scottish Environmental Protection Agency, East Kilbride 2003.
15. 67/548/EEC, Council Directive of 27 June 1967 on the approximation of laws, regulations and administrative provisions relating to the classification, packaging and labelling of dangerous substances. OJ, 1967, **L196**, 1–98.

16. The Chemicals (Hazard Information and Packaging for Supply) Regulations, Statutory Instrument, No. 1689, HMSO, London, 2002, ISBN 0-11-042419-0.
17. 1999/45/EC, Directive of the European Parliament and of the Council of 31 May 1999 concerning the approximation of the laws, regulations and administrative provisions of the Member States relating to the classification, packaging and labelling of dangerous preparations. OJ, 1999, **L200**, 1–68.
18. 91/155/EEC, Council Directive of 5 March 1991 defining and laying down the detailed arrangements for the system of specific information relating to dangerous preparations in implementation of Article 10 of Directive 88/379/EEC. OJ, **L76/91**, 1991, 1–21.
19. Globally Harmonised System of Classification and Labelling of Chemicals (GHS), United Nations Publications, New York, 2003, ISBN 92-1-216463-3.
20. 79/831/EEC, Council Directive of 18 September 1979 amending for the sixth time Directive 67/548/EEC on the approximation of the laws, regulations and administrative provisions relating to the classification, packaging and labelling of dangerous substances. OJ, 1979, **L259**, 10–28.
21. 92/32/EEC, Council Directive of 30 April 1992 amending for the seventh time Directive 67/548/EEC on the approximation of the laws, regulations and administrative provisions relating to the classification, packaging and labelling of dangerous substances. OJ, 1992, **L154**, 1–29.
22. Notification of New Substances Regulations, Statutory Instrument 1993, No. 3050, HMSO, London, 1993, ISBN 0-11-034278-X.
23. 793/93/EEC, Council Regulation of 23 March 1993 on the evaluation and control of risks of existing substances. OJ, 1993, **L84**, 1–75.
24. The Notification of Existing Substances (Enforcement) Regulations, Statutory Instrument 1994, No. 1806, HMSO, London, 1994, ISBN 0-11-044806-5.
25. Technical Guidance Document in Support of Commission Directive 93/67/EEC on Risk Assessment for New Notified Substances and Commission Regulation (EC) No. 1488/94 on Risk Assessment for Existing Substances. Office for Official Publications of the European Communities, Ispra, 1996, ISBN 92-827-8011-2.
26. European Union System for the Evaluation of Substances, European Chemicals Bureau, Joint Research Centre, Environment Institute, Ispra, Italy, 1997.
27. 76/769/EEC, Council Directive of 27 July 1976 on the approximation of the laws, regulations and administrative provisions of the Member States relating to restrictions on the marketing and use of certain dangerous substances and preparations. OJ,1976, **L262**, 201–203.
28. Convention for the Protection of the Marine Environment of the North-East Atlantic, OSPAR Commission, London, 1992.
29. 2003 Strategies of the OSPAR Commission for the Protection of the Marine Environment of the North-East Atlantic, OSPAR Commission, Reference No. 2003-21, OSPAR, New Court, 48 Carey Street, London WC2A 2JQ, 2003.
30. The Offshore Chemicals Regulations, Statutory Instrument 2002, No. 1355, HMSO, London, 2002, ISBN 0-11-039966-8.
31. 91/689/EEC, Council Directive of 12 December 1991 on hazardous waste, OJ, 1991, **L377**, 20–27.

32. The Special Waste Regulations, Statutory Instrument 1996, No. 972, HMSO, London, 1996, ISBN 0-11-054565-6.
33. Review of the Special Waste Regulations, Enviros Aspinwall report for the Department of the Environment, Transport and the Regions, 2000.
34. Hazardous Waste Regulations- interim guidance on premises notification, Defra Publications, Admail 6000, London Sw1A 2XX, 2005.
35. Stockholm Convention on Persistent Organic Pollutants, Depository Notification C.N.531.Treaties –96 of 19, 2001.
36. 850/2004/EEC, Regulation of the European Parliament and of the Council of 29 April 2004 on persistent organic pollutants amending Directive 79/117/EEC. OJ, 2004, **L299**, 5–21.
37. Rotterdam Convention on the Prior Informed Consent Procedure for Certain Hazardous Chemicals and Pesticides in International Trade, Conference of Plenipotentiaries, Rotterdam, The Netherlands,1998.
38. 304/2003/EEC, Regulation of the European Parliament and of the Council of 8 January 2003 concerning the export and import of dangerous chemicals. OJ, 2003, **L63**, 1–26.
39. Bahia Declaration on Chemical Safety, Final report of the Intergovernmental Forum on Chemical Safety, IFCS/ForumIII/23w, 2000.
40. Basel Convention on the Control of Transboundary Movements of Hazardous Wastes and their Disposal adopted from the Proceedings of the Conference of Plenipotentiaries of 22 March 1989, Secretariat of the Basel Convention, 13-15 Chemin des Anemones, CH-1291 Chatelaine, Geneva, Switzerland.
41. Sustainable Production and Consumption of Chemicals – a strategic approach, HMSO, London,1999, ISBN 1-851123-33-4.
42. Chemicals in Products: Safeguarding the Environment and Human Health, Royal Commission on Environmental Pollution. The Stationery Office, Norwich, 2003, ISBN 0-10-158272-2.

Disclaimer: Any opinions expressed in this article are entirely those of the author and should not be interpreted as representing government policy.

Chemicals Risk Assessment and Management

ELLIOT G. FINER

1 Introduction

This article is a personal and fairly sceptical tour round chemicals risk assessment and management. The views are my own, and may not be shared by organisations with which I have or had connections.

1.1 Chemicals in Modern Life

In considering how to mitigate the environmental and health effects of synthetic chemicals, the first question must be: could we reduce or eliminate our use of these products? I now devote some paragraphs to answering this.

The chemical industry in the UK has annual sales of £33 billion. Its products are used in almost every manufacturing activity. Our health, food, transport, shelter, communications and leisure activities depend on them.

Some chemicals end up in products which most of the general public easily recognise as chemical-based, being formulated mixtures of chemicals. Examples are medicines, fertilizers, pesticides, paints, inks, dyes, adhesives, cosmetics and toiletries, disinfectants, polishes, and cleaners. A second category is where the contribution of chemicals may be less obvious to some, but they still form a large part (or all) of the product, *e.g.* plastic ware, a pair of trainers, a car, a computer. A third category of chemicals is additives, where the chemicals may form only a small proportion of the product, but are essential to its properties, *e.g.* in rubber products and paper. A fourth use of chemicals is in production processes for non-chemical products, *e.g.* the etching of glass or the manufacture of silicon chips. Finally, there is a large market in chemicals whose purpose is to take part in chemical reactions to produce other chemicals.

Food production accounts for a huge use of chemicals. The challenge facing farmers is to produce enough food to feed a world population which has passed 6 billion and is forecast to reach 9 billion by the middle of this century. Crops need nitrogen, phosphorus and potassium from the soil. Even if these elements were there originally in adequate quantities, repeated cropping will exhaust the soil unless fertilizer is added. Of the three, of course only nitrogen can be incorporated from the air by plants, and then only in limited circumstances. There is no way in which the phosphorus which plants need for DNA and other essential biochemicals can be regenerated in soil other than by the addition of natural or man-made fertilizer.

Synthetic fertilizers have tremendous advantages over natural fertilizers such as manure, chief among which is the amount needed. To replace the 30 million tonnes of fertilizer used annually in Europe would require 2 billion tonnes of dung. This would require 1 million movements – of trucks – each day.

All crops are susceptible to disease and attack by pests. Those of us who have grown our own fruit and vegetables are only too familiar with rot, mould, attack by slugs, and attack by insects. Even so, we can barely imagine what it must be like to see these disasters happen to crops on which one depends for food, as distinct from those grown as a hobby. The Irish in 1845–1849 did not have to imagine it as they saw their potatoes destroyed by late blight. Over a million people died. Pests account for over 40% of pre-harvest losses of untreated key crops such as potatoes, maize, rice, and wheat.

For most people, foodstuffs arise in the wrong place or at the wrong time of year, or both. Preservation allows transport and storage. About 45% of the world's population now live in towns and cities, and it would be impossible to feed them without the use of chemicals in packaging and preservation. In India, up to 60% of every delivery of tomatoes transported is rotten before it reaches its destination.

Chemical preservation goes back millennia, including pickling, salting, smoking, and preservation in sugar and alcohol. These techniques combat growth of bacteria and fungi. Modern methods are similar, but now include widespread use of refrigeration and freezing, which of course use chemicals as refrigerants. Pests and oxidation are also kept at bay by plastic packaging and the use of antioxidants.

Health Improvement accounts for another key use of chemicals. We have made great strides in improving life expectancy and quality over the last century. In 1899, out of a population of around 35 million, 100,000 people died in England and Wales from infectious diseases alone. Life expectancy was about 42 years. Now, life expectancy at birth for a male in the UK is about 76 years.

If you ask people how chemicals improve health, their thoughts probably go first to pharmaceuticals – maybe antibiotics and analgesics. The availability of anaesthetics has revolutionised surgery, and of course there are many further types of pharmaceuticals in common use. But water purification and sewage treatment may have had an even bigger impact on health. Even today, the World Health Organisation estimates that 9 million people die each year from drinking water that has not been properly disinfected.

Chlorine has been the chemical that has made the biggest impact on drinking water quality, fulfilling the three most important requirements of a disinfectant: effectiveness, relatively low cost, and the ability to leave a residue in the distribution system to prevent the re-growth of microorganisms. Deaths from typhoid in the USA dropped from 150 to about 3 people per 100,000 population per year between the years 1900 and 1960, largely as a result of water chlorination. In the 1980s and 1990s, however, a key target of the major environmental NGOs was to get the industrial use of chlorine phased out. The Peruvian Government accordingly in the late 1980s started to remove chlorine from its country's water supply. This allowed a devastating cholera epidemic, which started in 1991, lasted till 1996, and spread across South America. It claimed more than 6000 lives and afflicted over 800,000 people.

Thus, to solve environmental and health problems arising from synthetic chemicals we cannot look to some idyllic world where synthetic chemicals are not used. If it is idyllic, it uses chemicals. We have instead to devise and implement ways of assessing and controlling the risks of making, transporting, and using those chemicals. And since we need the chemicals, we might as well enjoy the benefits of a thriving chemical industry – many interesting, well-paid jobs, and a substantial contribution to the wealth of the country. It would be both immoral and stupid to continue to use billions of pounds' worth of chemicals while enacting legislation which drove the manufacture of those chemicals to other parts of the world.

1.2 The Chemical Industry

The chemical industry (including pharmaceuticals) is one of the UK's biggest manufacturing sectors, and is the country's biggest manufacturing exporter. It employs about 250,000 people directly, and supports several hundred thousand jobs throughout the economy. It spends about £3 billion a year on new investment, and devotes about 9% of its sales income to R&D.

In 2003, the value of world chemicals production was €1619 billion. With sales of €556 billion, the EU (25) was the leading chemicals-producing area in the world. The EU chemical industry accounts for around 12% of EU manufacturing industry's gross value added. Over the 10 years to 2003, the EU chemical industry grew by an average 3.5% per annum, while growth in the US and Japan was 1.7%.

The total quantity of chemicals produced in 2003 worldwide was about 780 million tonnes. This includes 93 million tonnes of oxygen and nitrogen extracted from the atmosphere, some of which would eventually find its way back there and cause no pollution. The figure excludes, as best I could calculate, products (*e.g.* polymers) produced from other chemicals.

All this material ends up in the environment sooner or later. So in broad terms, the amount of chemicals produced each year, and ending up in the environment one way or another, is about ¾ billion tonnes, and growing. This is the scale of the issue. Of course all these chemicals come from the environment in the first place: around one-third of a billion tonnes is derived from oil and gas, with the rest coming from water, air and minerals. The environmental and health risks posed by manufactured chemicals are, however, obviously quite different from those of their source materials.

There is scope to reduce the risks posed by this volume of chemical products:

- through substitution of risky by less risky chemicals;
- by making better products (*e.g.* stronger plastic for shopping bags, thereby allowing less plastic to be used);

- by recycling (at present about 7% of the 3 million tonnes of plastic waste arising annually in the UK is recycled);
- and by designing 'greener' chemical processes, which use smaller quantities of chemicals and/or less risky chemicals and produce less by-product.

Some regulatory regimes aim to spur substitution, while commercial pressures and sponsored research will lead to better products and greener processes.

2 Types of Risk

Chemicals pose three types of risk: directly to health, directly to the environment, and to health *via* the environment.

2.1 Health Effects of Chemicals

The direct and immediate risks to health posed by some chemicals have been known for ages. Acids and alkalis corrode tissue, some chemicals explode or readily catch fire, and many chemicals, natural and synthetic, are acute poisons. In a sense, these risks are more easily controlled, and therefore are less of a problem and less feared, than risks which manifest themselves years or decades after exposure, or which afflict subsequent generations.

There are some health problems in this latter class which seem to have grown over the period of growth of the chemical industry. In particular, while most of the rise in cancer is attributable to the fact that we are living longer and therefore stand a greater chance of getting it, some increases in cancer cannot be totally thus explained. Cancers of concern include:

- testicular cancer, which increased in incidence by 55% between 1979 and 1991 in England and Wales;
- breast cancer, which has been estimated to have increased in incidence by 1% a year in the USA since the 1940s;
- prostate cancer, which increased by 40% between 1979 and 1991 in England and Wales, though at least some of this increase is due to improved diagnosis, as well as longer life.

There also seems to be a rise in asthma, and possibly in reproductive health problems additional to cancers.

These problems are certainly not necessarily caused by synthetic chemicals. There are many other possible factors, including:

- lifestyle, for example sitting in cars wearing Y-fronts; higher temperatures in homes and offices; reduced exposure to dirt in childhood; use of illegal drugs;
- air pollution from traffic;
- diet, including greatly increased consumption of soya, which contains powerful endocrine modulators, and increased consumption of coffee, which contains many natural carcinogens.

Chemicals Risk Assessment and Management 25

But we do know that some chemicals can cause cancers. Cancer of the scrotum was observed amongst chimney sweeps by Percival Potts in London in the 18th century, caused by exposure to benzo(a)pyrene and other chemicals present in soot. Vinyl chloride monomer can cause cancer of the liver, benzene can cause leukaemia, and beta-naphthylamine can cause bladder cancer. We take extreme care to avoid exposure of people to these chemicals. We do, however, still use them in controlled conditions, because society has judged that the benefits of doing so outweigh the risks once we have the right containment and safety measures in place. Probably the most important issue is what to do about assessing and managing the risks posed by chemicals which are used in uncontrolled conditions. I return in Section 4 to the evidence on whether synthetic chemicals contribute to disease in the general population.

2.2 *Environmental Effects of Chemicals*

During the last century, and in some cases earlier, pollution of air, water and land by emissions of chemicals from factories and transport was dreadful. Most of society had the idea that the environment could be treated as a free dump. The majority of these emissions were nothing to do with the chemical industry, but rather were produced by coal gas plants, iron and steel plants, electricity generators, trucks and cars, domestic coal fires, and others. Indeed, transport and electricity generation remain the world's great polluters, if we include CO_2.

Irrespective of who emitted the pollutants, we know what the results were for animal and plant health and biodiversity. Or do we? Certainly in extreme cases, such as dumping of chemicals in lakes and rivers, there has been damage, sometimes great, to the quantities and variety of wildlife and plant life. Certainly there can be great problems caused by eutrophication, which is an extreme increase in biological productivity in response to an increase in the concentration of nutrients such as nitrates and phosphates (normally from run-off from fields treated with fertilizers). Excess nutrient concentrations in freshwaters, coastal waters, or soils can result in the growth of organisms not usually dominant in the ecosystem.[1] Certainly there have been, and still are, local acute episodes or local longer-term poisoning ('direct toxicity') of plants and other life forms, caused by accidental or deliberate emissions from factories etc. Certainly it is established to the satisfaction of most observers that emissions of CO_2 are contributing to global warming; and we know that chlorofluorocarbons damaged the ozone layer. But it is surprisingly hard to find scientifically reported evidence of environmental harm caused by lower levels of pollution by industrial chemicals (I deal later with pesticides).

This is not to say that there are no such effects, and it is certainly not to say that pollution is not a problem, and should be permitted. I return below to the issue of what the problem really is.

2.3 *Health Effects of Chemicals in the Environment*

An example of an indirect health effect of chemical pollutants in the atmosphere is that asthma can be provoked by ozone created photochemically from oxides of nitrogen and volatile organic compounds emitted from vehicles and by industry. Another case was lead, which used to be added to petrol to help its combustion characteristics in car engines. The negative effects of lead emissions became visible only after cars became used in great numbers. Lead leads to cognitive retardation in children and causes learning disabilities, hearing loss, reduced attention span, and behavioural abnormalities as well as kidney damage. Cognitive and growth defects also may occur in infants whose mothers are exposed to lead during pregnancy. Once this became clear, lead in petrol was phased out in most developed countries, though not without opposition. It is still not phased out worldwide.

A major source of public worry is the perceived health effects of synthetic chemicals in food. Many consumers are prepared to pay extra for 'organic' food – food produced by farmers without the use or involvement of synthetic fertilizers, pesticides, and herbicides. It is estimated that the sales of organic food have gone up by about tenfold in the past 10 years. Some of consumers' worries are because of concerns about the toxic effects of pesticide residues. Sir John Krebs, chairman of the Food Standards Agency, said in a speech[2] in 2003:

> In our view the current scientific evidence does not show that organic food is any safer or more nutritious than conventionally produced food.
>
> Nor are we alone in this assessment. For instance, the French Food Safety Agency (AFSSA) has recently published a comprehensive 128-page review which concludes that there is no difference in terms of food safety and nutrition.
>
> Also, the Swedish National Food Administration's recent research report finds no nutritional benefits of organic food.

In fact organic farmers do use some pesticides: these derive from natural sources and include the insecticidal bacterium *Bacillus thuringiensis*, and botanically derived substances such as tree and plant oils, including pyrethrins. Pyrethrins are highly toxic to fish, and in high doses cause tumours in rats. Synthetic pesticides also pose acute toxicity risks if abused, and in the US each year about 67,000 pesticide poisonings, resulting in an estimated 27 accidental fatalities, are reported.[3]

3 Risks and Hazards

3.1 Risk Assessment

To assess the risk posed by a chemical, two sets of information are needed: the hazards posed by the chemical's intrinsic qualities, and the potential of the chemical to cause harm to people or the environment by interacting with them ('exposure'). In an ideal world, each of these is characterised and quantified, and risk reduction measures and regulations are applied accordingly.

If only life were that simple. In the real world, testing for all possible hazards requires numerous physical laboratory procedures, extensive use of laboratory animals (which raises serious ethical issues), and considerable elapsed time (to allow experimenters to observe long-term effects on plants, animals, or their offspring), and even then can only account for hazards which are already known. Nobody had thought of depletion of the ozone layer when chlorofluorocarbons were first marketed. Nobody had thought to test for teratogenic (interfering with normal embryonic development) effects when thalidomide was first sold.

In practice, hazard assessment regimes for chemicals do not demand a full *ab initio* assessment each time a new chemical is marketed. Maximum use is made of knowledge of the hazards posed by related chemicals.

For chemicals which find their way into public use, calculating exposure can be even more difficult than assessing hazards. A given chemical may be used in hundreds or even thousands of different applications. The manufacturer is unlikely to know what they all are, and indeed some applications may be deliberately kept secret by the users of the chemical since those uses may represent the users' sources of commercial competitive advantage. Where there is a supply chain of considerable length, the manufacturer may have little or no influence, let alone control, over how the chemical is finally used – what risk analyses are undertaken and what precautions are taken.

Nevertheless, society cannot be paralysed by the difficulties of risk assessment into not doing it. We need the chemicals, for the reasons outlined in the first section of this article, and so we need a way of minimising the risks of their doing damage.

One way to overcome this problem is to rely on hazard assessment alone. But a moment's thought shows that such an approach would rob us of some extremely useful chemicals. For example, car batteries contain sulphuric acid, at sufficient concentration to be very corrosive. On a simple hazard assessment, plastic boxes containing sulphuric acid would not be permitted to be sold to consumers. But in practice those robust plastic boxes – the battery cases – keep us from harm, and permit an extremely important use of the acid. Even so, there are some injuries each year caused by battery acid, and this demonstrates another point – that society has to accept a trade-off between risk and utility. We all do this without much thought in other areas of life, like transport, or even having a bath (the site of many domestic accidents)!

3.2 The Importance of Dose

The most famous saying in toxicology is 'Dosage alone makes the poison', written by Paracelsus, the 'father of toxicology', in the 16th century. Any substance is potentially toxic if the dose and duration of exposure are high enough. People die from drinking too much water and from eating too much salt.

There are many ways in which a chemical might affect the health of an organism. These include corrosive or irritant effects, acute and chronic toxicity, effects on the nervous system, impairment of the reproduction of cells or the organism (carcinogenicity, mutagenicity – *i.e.* causing genetic mutations – and reproductive toxicity), and damage to the hormone system (endocrine disruption). For each type of toxic effect, there are specific tests designed to determine whether the effect is evident at different levels of concentration or dose.

The key question relating to low levels of chemicals in the environment is how organisms react to doses much lower than those normally shown by such tests to cause harm. The answer comes from a dose-response (or, more precisely, a concentration–response) curve. This illustrates the relation between the amount of a chemical administered to an animal and the degree of response it produces. This response is measured by the percentage of the exposed population that shows the defined effect. If that effect is death, such a curve may be used to estimate an LD_{50} value. LD_{50} (Lethal Dose 50) is the dose of a chemical which kills 50% of a sample population. Such values are widely used as an effective measure of the potential toxicity of chemicals.

In very many systems, dose–response curves follow a standard shape, shown in Figure 1.

This is the shape expected when the response is due to binding to receptor sites. The interesting part of the curve for most chemicals which find their way into the general environment (as opposed to chemicals released as a result of an accident) is the bottom end. Modern analytical techniques are amazingly sensitive, and can detect extremely low concentrations. WWF, for example, reported in 2004 'Results from the widest ranging European survey of human toxic contamination show that 76 persistent, bio-accumulative, and toxic industrial chemicals were present in the blood of those tested.' These results included concentrations lower than 1ng/g blood, though some were much higher. How significant are these low concentrations?

A concentration of $1 ng\ g^{-1}$ blood equates to a total amount of about 4 µg in the blood of a man. We do know from our personal experience that some pharmaceuticals can have significant biological effects at low concentrations, but even so, we normally take much greater doses than this, generally in the range 1 mg to a few hundred milligrams. And pharmaceuticals are compounds designed to produce maximum (while safe) biological effects from minimum amounts. On the other hand, the dose of Botulinum toxin A which kills humans is about 1 µg. Furthermore, when considering safety we have to think about the possible effects of chemicals on the weakest members of society, *e.g.* unborn children.

Most regulatory agencies assume that dose-response curves are all broadly of the shape shown in Figure 1 in that there is a dose below which there are no significant biological effects. This accords with observation, and also with expectations of how organisms should react when a response is due to binding at a receptor site. Furthermore, many biologists believe evolutionary pressures have caused organisms

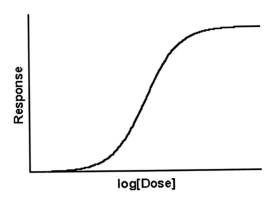

Figure 1 *Shape of typical dose–response curve*

to develop mechanisms to deal with low levels of insult. On the other hand, some environmentalists believe that this may not be the case: for example, WWF say[4] 'it is likely that many endocrine-disrupting substances can cause effects at lower concentrations than currently supposed, and indeed some of these substances may exhibit inverted U-shaped dose–response curves'.

Finally, there is the issue of possible synergistic effects of mixtures of chemicals. Is it legitimate to consider a separate dose–response curve for each chemical in a mixture to which an organism is exposed? Or might the chemicals interact to reinforce each other's harmful effects, or even to negate each other's effects? We do know that pharmaceuticals can interact with each other in some cases and produce harmful or enhanced side-effects. In its comments to the Royal Commission on Environmental Pollution during the scoping of its chemicals study,[1] the Natural Environment Research Council said 'Despite much discussion our actual knowledge of the synergistic effects of exposure to various groups of chemicals remains poor'.

4 The Evidence for Harm Caused by Industrial Chemicals

It is clear that huge quantities of man-made chemicals find their way into the environment, and that there are many pathways by which synthetic chemicals may get into organisms, including ourselves. It is clear that some chemicals can be acutely toxic to a range of organisms, and that some can produce harmful effects which may not become apparent for decades. What is not clear is whether harm is being done to organisms at the doses of these chemicals that they receive. What value would society obtain in return for spending billions of pounds on reducing or eliminating these doses and accepting countless lost opportunities for benefit from chemical products?

4.1 Disease

Bjorn Lomborg addressed some of these issues in his controversial book *The Skeptical Environmentalist*.[5] He used data from reputable sources such as the World Health Organisation to show that US cancer death rates, when adjusted for changes in life expectancy and the prevalence of smoking, have steadily declined over the last 50 years, by about 30% over the period. Changes in breast cancer rates can be explained by increases in the age when women have children, decreases in the number of children they bear, and by increase in body weight. Other cancer rates show complex changes over the period, but in some cases there have been dramatic declines – *e.g.* women have experienced a decline in cancers of the uterus by 81%, and stomach cancer rates have declined even more, for men and women. This last change is most likely to be mainly a result of improvements in diet (more fresh fruit and vegetables).

If low-level exposure to chemicals causes disease, then one might expect that professional chemists would die young. In fact the reverse is true. Although diseases that are statistically significant in terms of chemist deaths were identified, a retrospective survey[8] carried out by the Royal Society of Chemistry for the years

1965–1989 found that professional chemists tended to live longer than other workers in the same social groups.

This study determined the causes of mortality of 4012 chemists among 14,884 members of The Royal Society of Chemistry who were followed for a period of 25 years. The results were consistent with other studies. Overall, the results showed that chemists had a relatively low mortality from common killer diseases in general, but that the incidence of death due to cancer was disproportionately higher. The cancer tended to be lymphomas and pancreatic cancers, and leukaemia and intestinal cancers. There were also, however, clear correlations observed for type and incidence of cancer with chemist location. For example, death due to lymphatic cancers was observed to be much higher than expected in the case of Swedish chemists. The prevalence of other cancers, such as testicular, and of certain heart diseases, was found to be relatively high for chemists.

The good news (for us chemists) is that if all causes of death are combined, then chemists have a relatively low-mortality rate. We appear to be much less susceptible than other groups to respiratory and heart diseases, and to accidents (I suppose that the last fact is due to our health and safety training). Of the people studied who died, 75% were alive at the age of 63, 50% at the age of 73, and 25% at the age of 80. To make the results more meaningful, other facets of lifestyle were also taken into consideration. It is significant that there were fewer smokers amongst chemists than there were for several other groups of professionals.

According to the European Environment Agency,[6] it is very difficult to identify any cause–effect relationships between the most common diseases in the EU and the factors which may have given rise to those diseases, including genetics, lifestyle, and environment. In the face of this 'considerable uncertainty', the Agency recommends in many cases the use of the 'precautionary principle', as do many other bodies.

4.2 The Precautionary Principle

The precautionary principle says: 'Where there are threats of serious or irreversible damage, lack of full scientific certainty shall not be used as a reason for postponing cost-effective measures to prevent environmental degradation'. This wording comes from the 1992 Rio Declaration on the Environment and Development. Note the inclusion of the phrase 'cost-effective', which puts the onus on proponents of measures to have identified costs associated with the harm the measures are designed to tackle (though some environmentally beneficial measures are intrinsically cost-effective, *e.g.* many energy efficiency investments).

4.3 Pesticides

What about harm to plants, insects, birds, fish, *etc.* from synthetic chemicals in the environment? The first major impetus to such concerns came from Rachel Carson's book[7] *Silent Spring*, published in 1962. She wrote that pesticides such as DDT were spoiling the earth, potentially leaving us with a silent spring, devoid of singing birds. Among other things, the book claimed that DDT interferes with bird reproduction and causes cancer in humans. After its publication the chemical was linked to the thinning of eggshells in some avian species. The US Environmental Protection Agency was created in 1970, in no small part due to *Silent Spring*, and two years later DDT became the first chemical it banned. Most other industrialised nations followed suit, and developing countries were pressurised to do the same.

We now know that, while DDT is highly toxic to insects and fish, very large doses are needed for it to poison other animals, including birds and humans. It does not seem to cause cancer, nor does it seem to be the cause of eggshell thinning.

On the other hand, DDT is demonstrably effective at controlling the mosquitoes and other insects that transmit malaria and typhus. Thanks principally to DDT, in the years after World War II malaria was eradicated in the US and sharply reduced in many tropical countries. Venezuela recorded eight million cases of malaria in 1943; by 1958 that number was down to 800. The World Health Organization estimates that DDT has saved 50–100 million lives from malaria prevention alone. In recent years, however, the disease has staged a comeback. Globally it quadrupled during the 1990s. It now kills well over a million people a year, and is a major factor in reducing the GDP of afflicted countries. The resurgence of malaria is due to a variety of factors, including changes in land use and possibly climate, but some experts say the phasing out of DDT is one of the causes.

Even if Rachel Carson was wrong about the risks of using DDT, the movement which she started, towards much greater care in the use of pesticides, has undoubtedly been beneficial. There are now stringent controls over the use of insecticides and other pesticides in the EU and other parts of the world. In the UK, these are enacted in the Food and Environment Protection Act 1985 and the Control of Pesticides Regulations 1986. Section 16 of the Act describes the aims of the controls as being to protect the health of human beings, creatures and plants; safeguard the environment; secure safe, efficient and humane methods of controlling pests; and make information about pesticides available to the public.

There are still many points of view about the risks involved in the use of pesticides, and arguments about the safety of individual pesticides or classes of pesticide. But there are not many people who argue for a drastic tightening of the overall regulatory regime governing the use of pesticides in the UK. The remaining big issue is the overall regulatory regime governing the marketing and use of chemicals in other products, of the sort mentioned in the first section of this article. There is also an issue about the environmental effects of pharmaceuticals for human use. These are potent substances which may be excreted unchanged or as still potent breakdown products.

4.4 *Endocrine Disrupters*

The class of chemical which has received most attention from environmentalists in recent years is endocrine disrupters. These are substances which cause adverse effects in an organism or its progeny by interacting with hormonal (endocrine) pathways, especially those involved in reproductive and thyroid gland function.

Some naturally occurring endocrine disrupting substances have been known for ages. For example, red clover contains oestrogens, which can render grazing sheep infertile, less sexually receptive, and more aggressive. Clover oestrogens (and soya oestrogens when sheep forage on that plant) can cause ovarian cysts and irreversible endometriosis in ewes, and blocked urethra, enlarged teats and lactation in wethers (castrated rams).

Synthetic chemicals which have been shown to have endocrine disruptive properties in the laboratory include bisphenolic and alkylphenolic compounds, and tributyl tin, which was widely used as an anti-fouling biocide and as a fungicide in paints. Its use is now restricted.

According to the Royal Commission on Environmental Pollution,[1] studies of environmental endocrine disrupters have illustrated sub-cellular events with clear implications for reproductive performance, but only in a few cases has it been possible to link biochemical effects of a single compound to significant changes at the population level. Such effects have been observed in fish exposed to sewage effluent, but these are believed to be due to steroidal oestrogens in the effluent.

4.5 Synthetic versus Natural Nastiness

The truth is that it seems so obvious to environmentalists that it is wrong to allow man-made chemicals to pervade the environment that many do not consider it really necessary to demonstrate that those chemicals do real damage, though environmental lobbyists do publicise ways in which industrial chemicals could potentially, in theory, cause harm. Indeed, most of us instinctively react in favour of better controls when we learn, for example, that industrial chemicals are detectable in the environment in polar regions. We feel that such regions should be pristine, and we consider industrial chemicals to be sullying that purity.

This feeling is related to an instinctive belief, held by many, that synthetic chemicals are somehow different from naturally occurring substances. Greenpeace uses the following on the front page of the chemicals (called 'toxics') section of its website: 'Unborn babies are exposed in the womb to synthetic chemicals.' The implication is that synthetic chemicals are by definition harmful.

Up to the 1830s people thought that natural chemicals were different from chemicals made in the laboratory. It was believed that organic substances could only be formed under the influence of the 'vital force' in the bodies of animals and plants. It was a German chemist, Wohler, who proved this idea false, by synthesising urea from inorganic materials in 1828. He thereby showed that a compound known to be produced only (till then) by biological organisms could be produced in a laboratory, under controlled conditions, from inanimate matter. This *in vitro* synthesis of organic matter disproved the common theory (vitalism) about the *vis vitalis*, a transcendent "life force" needed for producing organic compounds.

The fact is, of course, that natural substances are identical in every respect to the same substances made in the laboratory or chemical factory. But I suspect that the idea of a vital force lingers in some people's minds. I return in Section 6 to the influence of innate beliefs.

Some naturally occurring substances are amongst the most poisonous materials known. I noted earlier that the dose of botulinum toxin lethal to a man is about 1μg. One milligram of saxitoxin, an algal toxin, will kill you; the same is true for tetrodotoxin, present in puffer fish. By contrast the fatal dose of hydrogen cyanide is 50 mg, and a fatal dose of paraquat is 1400 mg.

A common belief is that man has evolved to cope with naturally occurring chemicals, but not with those which do not occur naturally. This is certainly not universally true. The potato (*Solanum tuberosum*), for example, is well known to contain a range of potentially toxic chemicals, including carbohydrate derivatives of 3-hydroxysteroidal alkaloids, also known as glycoalkaloids. In cultivated potatoes, the major glycoalkaloids are α-solanine and α-chaconine, which are derivatives of solanidine but with different carbohydrate groups attached to the molecule. These have an LD_{50} of 42 mg kg^{-1} (mice) and 84 mg kg^{-1} (rats) respectively. Particularly high levels of these compounds are found in potatoes that are green owing to exposure to light. Their presence is, of course, natural – probably part of the plant's defence mechanisms against pathogens – and is therefore independent of whether the crop was produced by organic farming.

Humans have measurable levels of solanidine in their blood, from potatoes in the diet. Amounts normally range from 0.3 to 5 µg per litre, which is the same order of magnitude as the levels of some industrial chemicals which environmental groups report in news releases. Accidental poisoning by potatoes unusually rich in these toxic glycoalkaloids is not uncommon, and in 1984 Morris and Lee[10] wrote a report of well-documented incidents involving 2000 people, many of whom were hospitalised and 30 of whom sadly died.

4.6 Conclusions about the Need for Tighter Regulation

My personal view is that society has reached a stage where it should no longer be acceptable to allow large quantities of industrial chemicals to find their way into the environment without our knowing as best we can the implications of that process for the health of man and other organisms, and so being able to prevent the process where appropriate. The lack of evidence of actual harm from low levels of synthetic chemicals in the environment may be due to the difficulty and expense (and therefore lack of) the large-scale epidemiological studies which would probably be needed. Even where there is no known risk to health or the environment, I think there is an aesthetic argument for keeping our environment clean.

Although I am not concerned in principle by the fact that skilled analysts using modern technology may be able to find minute traces of industrial chemicals in my blood, I want to know that those chemicals pose no risks at the concentrations found. Furthermore, while I reject as scare stories the large number of potential (but untested) pathways for industrial chemicals to do harm which some environmentalists dream up, I am certain that we do not know everything about possible pathways to harm: and so in principle I am unhappy about significant concentrations of industrial chemicals lingering in the environment. I am particularly concerned about pollutants which are persistent, bio-accumulative, and already known to be highly toxic, and only slightly less concerned about those which are persistent and bio-accumulative but have not (yet) been shown to be highly toxic.

I do not, however, believe that we must clean up the environment at all costs. More than lip service must be paid to weighing costs and benefits.

5 Cost–Benefit

Any chemical control regime imposes significant costs on the industries it regulates and on the society it protects. Companies in the regulated sectors have to bear the costs of testing the products, and also the costs caused by delays in putting the products on the market while the tests are carried out and evaluated. One might, however, think this is a price worth paying for some assurance that risks associated with the product can be managed. Perhaps even more damaging is uncertainty over whether a new product can be put on the market at all. This is a risk routinely borne by the pharmaceutical industry, most of whose potential products in fact do not make it to the marketplace. One result of this poor success rate is that the development costs which have to be recouped from sales of those products which do make it to the market are very high. Another result is that repeated marketing of new pharmaceuticals is really only possible for giant companies.

The uncertainty faced by companies is even greater if the regulatory regime involves comparisons with other products, which may or may not themselves yet be on the market. This is true of the EU Biocidal Products Directive (98/8/EC). This directive introduced the concept of comparative risk assessment, whereby substances can be prevented from being marketed, including being withdrawn from the market, if there is judged to be an alternative substance that presents a lower risk to human health or the environment. So a company might spend millions of pounds developing a new biocide, spend further large sums to get it tested, and put it on the market, only to see it forcibly withdrawn because a new product is being marketed, a product which EU committees have decided does the same job at lower risk.

You might think that it is a normal risk of business that a product fails because the market prefers an alternative product. That is true, but the Biocidal Products Directive imposes an additional risk, of an all-or-nothing nature and seen by many industrialists as potentially subject to political interference, since decisions are made by EU committees. Some argue that fewer new products will be developed because of these real and perceived risks. Others argue that the withdrawal of existing products will act as a spur to innovation to produce replacement products. My own judgment is that innovation in this sector will be reduced, with resultant lack of consumer choice and probably poorer pest control and more damage to the environment. Niche products, which do a particular biocidal job very well but serve a limited market, are likely to suffer most, since there are smaller volumes over which to spread the costs of testing.

If a regulatory regime impedes innovation it will have particularly pernicious effects in the long term. If European regulation makes it more costly or uncertain to innovate in Europe than in other parts of the world, research and eventually company headquarters and manufacturing plant will migrate to where the research is easier. And it is not just a matter of international competitiveness. Constant innovation is essential for any product involved in fighting microbes, because even if we do not innovate, they will. Bacteria and viruses continually evolve to develop resistance to our chemical attacks on them.

In its initial proposals[9] for a new regulatory regime for chemicals – 'REACH', standing for Registration, Evaluation and Authorization of CHemicals (discussed in more detail in Section 9), the European Commission made little attempt to identify the harm caused by the chemicals it proposed to regulate, or to carry out a cost–benefit analysis. To justify introducing measures costing billions of Euros, the White Paper said:

> ... certain chemicals have caused serious damage to human health resulting in suffering and premature death and to the environment. Well-known examples amongst many are asbestos, which is known to cause lung cancer and mesothelioma or benzene which leads to leukaemia. Abundant use of DDT led to reproductive disorders in birds. Though these substances have been totally banned or subjected to other controls, measures were not taken until after the damage was done because knowledge about the adverse impacts of these chemicals was not available before they were used in large quantities.
>
> The incidence of some diseases, *e.g.* testicular cancer in young men and allergies, has increased significantly over the last decades. While the underlying reasons for this have not yet been identified, there is justified concern that certain chemicals play a causative role for allergies. According to the Scientific Committee on Toxicity, Ecotoxicity and the Environment of the Commission (CSTEE), links have been reported between reproductive and developmental effects and endocrine disrupting substances in wildlife populations.

Asbestos is not, of course, a chemical product.

The Commission subsequently did more work on costs and benefits. They estimated that the overall costs of the proposed REACH regime to the chemicals industry and its downstream users would be €2.3 billion and €2.8–5.2 billion respectively. Other studies put the cost to the chemical industry at €3.5–4 billion. The chemical industry expects that these costs will fall most heavily on small-and medium-sized enterprises and some speciality chemicals companies. Such companies typically manufacture many low-volume products. They are likely to have to undertake assessments as individual producers rather than as an industry sector or product group. Furthermore, they generally have smaller markets and so less opportunity to recoup costs.

On the basis of some fairly heroic assumptions, the Commission estimated the value of the health and environmental benefits of REACH to be around €50 billion over 30 years, thus hugely dwarfing the costs. That calculation has, however, been challenged[11] by Kramer et al.; these authors point out some fundamental flaws in the Commission's calculations, not least that the figures used by the Commission for estimating lives saved come from studies largely of the results of pesticide misuse. REACH does not deal with pesticides, which have their own regulatory regime, as explained earlier.

Even if the cost is at the top end of the ranges quoted above, the total cost per resident of the EU per year, over just 10 years, would be only €2. If EU residents were asked if they were willing to pay €2 each a year over 10 years for a system like REACH to operate, I believe that the majority would say yes. I therefore conclude that REACH cannot be dismissed on cost-benefit grounds. Nevertheless, every sinew should be bent to minimise costs.

6 Perception of Chemical Risks and the Roles of the Advocates

The political and legislative agenda on chemicals risk management has largely been set as a result of lobbying by activists, sometimes hugely aided by chemical industry incidents such as Bhopal. The Bhopal disaster of 1984 was caused by the accidental release of 40 tonnes of methyl isocyanate from a Union Carbide pesticide plant located in the heart of the city of Bhopal, in the Indian state of Madhya Pradesh. The leak killed thousands outright and injured between 150,000 and 600,000 others, at least 15,000 of whom died later from their injuries. Some sources give much higher fatality figures.

On the environmental side the main groups in the UK have been Greenpeace, Friends of the Earth, and WWF. Greenpeace's arguments about synthetic chemicals are centred around their possible impact on human health. They conduct a 'toxics campaign', which features in their statement of their overall aims:

> Greenpeace organises public campaigns
> - for the protection of oceans and ancient forest
> - for the phasing-out of fossil fuels and the promotion of renewable energies in order to stop climate change
> - **for the elimination of toxic chemicals** *(my emphasis)*
> - against the release of genetically modified organisms into nature
> - for nuclear disarmament and an end to nuclear contamination.

Greenpeace's mission statement is 'to ensure the ability of the earth to nurture life in all its diversity'. What would flow naturally from this mission statement is a campaign on chemicals which majored on the alleged damage synthetic chemicals were doing to the environment. Since in fact their campaign concentrates on potential threats to human health, a cynic might conclude that Greenpeace's work in this area is motivated by their being anti-synthetic chemicals, rather than being pro-environment. Furthermore, if their main concern about traces of chemicals found in the body were to do with the protection of people from harm, they would be campaigning for potatoes to be banned or genetically modified so as to produce lower levels of solanidine, which is detectable in the blood and which we know does actually kill people (see Section 4 above). So it is the synthetic aspect of synthetic chemicals which they are against.

In their 'safer chemicals' campaign, Friends of the Earth also deal exclusively with potential effects on human health. The objectives of Friends of the Earth Trust Ltd are: 'Friends of the Earth Trust is committed to the conservation, protection and improvement of the environment. It undertakes research, education and publishing and provides an information service on environmental problems and their solutions.' One might expect from this that their chemicals campaign would deal at least in part with effects on the environment. WWF's chemicals work also concentrates on potential human health risks from synthetic chemicals, but does as well address environmental issues.

Why are these groups so concerned about synthetic chemicals? At least part of the answer is to do with instinctive feelings, discussed in Section 4 above, that synthetic substances are somehow worse than naturally occurring substances and that natural places (and bodies) should not be sullied by the presence of synthetic materials. I believe that these instincts come first, and that people driven by them then construct arguments to justify action. We can see this process at work in lobbying about genetically modified organisms, where the objective is to ban them but the reasons adduced vary over time and from one person to another, from the creation of superweeds to fears about reduced biodiversity to possible allergic reactions.

Instinctive feelings are often to be trusted. The reason that some fears are instinctive is that they have become part of our genetic inheritance because they were so well founded that they aided survival. Most cognitive scientists believe that we are equipped with a number of different kinds of 'core intuition' and cognitive faculties (which appear to be computed in partly distinct sets of networks in the brain).[12] These were suitable for analysing the world in which we evolved. One is

> an intuitive version of biology or natural history, which we use to understand the living world. Its core intuition is that living things house a hidden essence that gives them their form and powers and drives their growth and bodily functions.

Another is a system for assessing contamination, coupled with the emotion of disgust.

These, then, are the innate instincts and emotions, as well as facts, against which lobbyists for the chemical industry have to fight. No wonder they have a hard time. Even when the industry has cleaned up its act, the public often does not realise that things have changed. Perceptions are rooted in the past. Furthermore, because of its past behaviour – which often was not out of line with expectations of the time – people developed mistrust for the chemical industry. So, even when the industry behaves well and tries to convince its neighbours and other that it is so doing, it is not always believed. Against this background, it is difficult for a politician to refuse to enact or endorse legislation designed to reduce the risks from chemical factories and products.

Parallels can be drawn with the factors which led to the establishment of the Food Standards Agency (FSA) in the UK. The FSA was established in 2000 after a number of food scares, including BSE. There was a mood of suspicion about the way that Government and the food industry handled food safety. The paradigm had been to assert that food was 'absolutely safe' and that 'scientists knew', and to 'decide, announce and defend'. The FSA believes that it has changed that: its messages are 'life is not risk-free', and 'although we turn to science, there is often no clear-cut answer'. It seeks to involve stakeholders early, and be open about the fact that decisions are essentially judgments.

The chemical industry has now for some years been following a similar path to that taken by the FSA. This approach recognises that societal attitudes have changed: people are now less likely to accept what they are told, and more likely to stick up for their rights.

Over the years, the regulation of chemical factories and products has become tougher and tougher. Where industry has fought against the principle of these regulatory regimes it has lost (though it may have delayed their introduction); where it has scored successes, however, is in making the regimes workable.

7 The Problems in Controlling Risks from Chemicals

The fate of a synthetic chemical might be any or all of the following:

- It is used neat by the final consumer. Not many chemicals are used neat by the general public, but some are used by non-chemical industry, *e.g.* acids for etching.
- It is sold as part of a formulated product, *i.e.* mixed with other substances but not chemically reacted. This applies to polymers and other components that are added to a polymer mix before the polymer is moulded; to inks and dyes; to solvents; to active and other ingredients of pharmaceutical products; to food additives; to detergents and other hygiene products; and to many other product areas.
- It is reacted with other chemicals before being available to the environment. This obviously applies to chemical intermediates, and indeed many chemicals never see the light of day, being produced in one chemical plant solely for use in others. It also applies to chemicals introduced into mixtures to change the properties of major components, *e.g.* chemicals which are added to rubber during the manufacture of tyres.

It is thus clear that controlling a chemical on the basis only of its intrinsic hazards would be nonsensical. A chemical which may become part of a formulated product which we put on our skin, spray on the land, use in DIY jobs in the home, or eat, needs a different regime from one which only leaves a chemical site if it is transported in sealed tankers to other chemical sites. And chemicals which become unavailable to the environment because they have reacted within the final product also need a different regime.

Because of this complexity, thought has to be given to how a chemicals control regime applies to imported goods. Since the aim is to protect the public and the environment, there should be no distinction between chemicals in products manufactured within the controlled territory (*e.g.* the EU) and those in imported products. It is relatively straightforward to achieve this aim for neat chemicals, which can be subjected at the point of import to the same regime as domestically-produced chemicals. REACH aims to do that. The same applies to formulated products. But what about chemicals in other manufactured products?

As an example, consider the computer on which I am composing this article. It contains a large number of chemicals – in the polymers making up the keyboard, in the display screen, in the cables, in the electronic components. Suppose identical display screens were made both in the EU and in the Far East. Would an EU chemicals safety regime apply to the imported screen, and if so, how? If not, why apply it to the screen manufactured in the EU, since both screens would pose an equal risk? Furthermore, if EU manufacturers had to bear and charge for the costs of testing while foreign manufacturers did not, consumers would quickly switch their purchasing to imported screens, with a resultant loss of jobs in the EU and no environmental or health benefits.

I do not think there is any straightforward answer to this problem. The best solution would be for the EU to press for adoption of its chemicals safety regime, or regimes of similar effect, by other regions of the world. Indeed, I expect that this will eventually happen as a result of pressure by the public in other regions – no region's politicians would wish to be seen to offer less protection for their citizens than that offered by another major region. However, there are two major caveats: first, this process may take considerable time, allowing distortion of trade in the meanwhile; and second, competing regions may be cleverer than the EU in devising regimes which offer protection while costing less and hindering innovation less.

8 Industry Initiatives

When general concern first arose about the health and environmental effects of the activities of the chemical industry, it was focused on the industry's manufacturing plant. People were concerned about accidental discharges, the health and safety of workers, and the results of planned discharges into waterways and the air. High profile accidents heightened concerns, and the highest was the Bhopal disaster, mentioned above, in 1984.

In 1985, the chemical industry in Canada conceived a voluntary initiative called Responsible Care, to address public concerns about the manufacture, distribution, and use of chemicals. Responsible Care was gradually adopted by other national chemical industry associations, and the number of countries now involved is 47, according to the International Council of Chemical Associations (ICCA).[13]

Under Responsible Care, companies in the chemical industry commit themselves to continual improvement in all aspects of health, safety, and environmental performance and to open communication about their activities and achievements. National industry associations are responsible for the detailed implementation of the initiative in their countries. Through the sharing of information and systems of checklists, performance indicators and verification procedures, Responsible Care enables the industry to demonstrate how it has improved over the years and to develop policies for further improvement.

There is no doubt that the chemical industry has been very successful in reducing the impact of its factories on the environment and human health. Many indicators show considerable improvements over the last couple of decades. Concerns of the general public now tend to concentrate on the impact of the industry's products rather than its plant, though neighbours of chemical factories still have many concerns.

But how much of the improvements in performance has been the result of voluntary action rather than tighter regulation? The regulation of chemical plant has greatly tightened over the period of the improvement. My view is that regulation and the anticipation of it has been the main driving force, but Responsible Care has been a great contributor. It has provided a focus for the industry to think of itself as taking positive steps rather than always being on the defensive, has given it improvement figures to be proud of, and has provided mechanisms for information sharing and collaboration in the development of improvement measures. I do not think that the industry would have got to where it is on plant safety and environmental performance so fast and so well without Responsible Care.

So could industry initiatives also contribute significantly to solving concerns about chemical products? In 1998, the ICCA established a programme to speed up testing of 'high production volume' (HPV) chemicals. These are defined as chemicals which are produced in volumes of more than 1000 tonnes a year in more than one region of the world. The ICCA did this because it was conscious of growing public concern, stimulated by environmental organisations, about the slow pace of existing work under an OECD programme to screen HPV chemicals. Only about 100 substances had been screened in the decade since the OECD programme had been established. In the absence of definitive sets of data, various countries were taking unilateral action, including publishing lists of 'unwanted chemicals'.

The initial ICCA list of HPV chemicals included about 1000 substances. The aim was to complete 'SIDS' (screening information data sets) hazard assessments for these priority chemicals by the end of 2004. ICCA established a public website where data would be published and progress could be monitored. Companies were asked to make promptly available any significant, new information regarding adverse health or environmental effects of the chemicals they were testing.

Initial progress was slow, for a number of reasons, including differences in view between regions of the world on the detail of some of the tests needed, difficulties in getting companies to collaborate, and difficulties in getting national governments to play their part in sponsoring submissions of dossiers to the OECD. In February 2005, according to the ICCA HPV initiative website, 287 chemicals had gone through the whole process, with a further 44 having completed a draft initial assessment. Thus although progress was faster than it had been before the ICCA initiative, it was considerably slower than the industry's self-imposed target.

There are greater problems in getting inter-company cooperation on product testing than there are in getting such cooperation on process safety and environmental measures. The former involves commercial managers, steeped in the need to compete; is expensive with no immediate obvious return; and may indeed come up with information which would see markets destroyed and profit targets threatened. The latter normally involves plant managers who are used to cooperating with opposite numbers from other companies, and whose work rarely is highly commercially sensitive. It is therefore easy to see why environmental groups have been calling for legislation to get products tested.

The UK chemical industry has for some time recognised the concerns about its products. The Chemical Industries Association has a voluntary 'Confidence in Chemicals' initiative, encompassing the HPV programme, dialogue with stakeholders, and improved stewardship of its products. In addition, it is attempting to improve relationships within the chemicals supply chain. It is working towards 'partnerships with organisations which represent manufacturers, formulators, distributors, resellers, hauliers and others involved in the marketing of chemicals help to promote continual improvement in health, safety and environmental performance through the product supply chain'.

9 REACH

The gap in chemicals regulation concerns industrial chemicals which were already on the market ('existing') in 1981, before an EU directive, on 'new' chemicals, came into force. Chemicals for sensitive uses – plant protection, biocidal, veterinary, cosmetics, pharmaceuticals, food additives, and animal feedstuffs – are covered by specific EU directives.

The proposed single regulatory framework would replace the current dual system for assessing risks of existing and new substances. Some 30,000 substances would be assessed through the REACH process. One major change from the previous regime would be that the duty to test and risk-assess chemicals would fall on industry rather than the authorities. The main elements are:

Registration

- Substances produced or imported in quantities of one tonne or more per year would have to be registered in a central database. The substances of highest concern would be set the shortest deadlines.
- Registration would not be required for substances produced and imported in quantities less than one tonne, or for substances for use in research activities, or for polymers.
- The amount of information required would be proportional to production volumes and risks.

Evaluation

- Dossiers would be evaluated for completeness.
- Member States' competent authorities would carry out evaluations of substances when they had reason to suspect a risk to human health or the environment.

Authorisation

- Authorisation would be required for highly problematic substances – CMRs (carcinogenic, mutagenic, or toxic to reproduction), PBTs (persistent, bio-accumulative, and toxic), vPvBs (very persistent and very bio-accumulative), and other substances with serious and irreversible effects on humans and the environment.
- Authorisation would be granted for these substances if risks could be adequately controlled or on socio-economic grounds if there were no technological alternatives.

Agency

- A European Chemicals Agency would be established to manage the registration database; it would also play a role in evaluation and authorisation. The Commission had foreseen that this Agency would be based in the Joint Research Centre in Ispra, but the Brussels European Council of 12 December 2004 agreed that it would be based in Helsinki.

The European Parliament is expected to examine the draft REACH regulation in the second half of 2005. Following that, a slightly modified version of REACH is expected to be tabled by the Commission. The Council will then become formally involved.

For the chemicals industry and its downstream users (*e.g.* carmakers who use thousands of chemical substances when producing cars), the Commission's plans conjure up fears of bureaucracy, lack of flexibility, loss of competitiveness, and job losses. Industry claims that the proposed system is too costly, unworkable, and would stifle innovation. In September 2003, the leaders of the UK, Germany, and France wrote an unusually condemnatory letter to the Commission expressing severe concerns about the competitiveness implications of a number of aspects of the draft legislation. On the other hand, the environmental movement and consumers' associations have accused the Commission of giving in to heavy lobbying from the industry and from the United States. According to the European Environmental Bureau, the proposed REACH system has several "loopholes" and "flawed approaches".

Specific proposals by industry to modify REACH include incorporating risk-based priority-setting, with greater use of information about exposure; changing the approach to chemicals in articles, because of fears of unfair competition from non-EU competitors (as discussed in Section 7 above); and applying restrictions and not authorisation as the preferred approach for risk management for chemicals whose use has to be limited.

The UK's Royal Commission on Environmental Pollution[1] (RCEP) also saw flaws in the REACH proposals, which it described as cumbersome and time-consuming. It felt that the core objective of chemicals policy should be to drive the progressive substitution of chemicals by chemicals of lower hazard or by non-chemical alternatives. It recommended

- listing all chemicals on the UK market
- sorting them according to simple screening criteria reflecting hazard and exposure
- evaluating selected chemicals identified by the screening process, or taking immediate action where danger is spotted
- taking risk management action quickly
- making information available throughout the supply chain.

Sensible though these proposals may seem, and indeed in line with some of industry's wishes, the EU juggernaut is rolling and will be impossible to stop. The RCEP's proposals may however inform and influence the UK government's negotiators as the REACH regime is finalised.

10 Conclusions

1. Synthetic chemicals are essential to modern life. We depend on them for quantity and quality of food, for healthcare and hygiene, for communications, transport and energy supply. Eliminating the risks by eliminating the products is not an option, though innovation will in some cases lead to greener products and processes.
2. Since we need the chemicals, we should nurture the industry which makes them. That industry provides large numbers of good jobs and creates considerable wealth. A risk control regime which drove the industry to other parts of the world would be immoral and self-harming.

3. Synthetic chemicals have caused health and environmental problems in the past. They may still be causing such problems, though there is very little evidence of actual current harm arising from synthetic chemicals in the environment, except when there are accidental or deliberate releases at relatively high concentrations. The lack of evidence may be due to the difficulty and expense (and therefore the lack of) the large-scale long-term epidemiological studies which might be required to demonstrate harm.
4. Some of the environmental groups' campaigns against synthetic chemicals seem somewhat at odds with the main agendas of those groups. They probably arise from instinctive feelings, shared by most of us, against the sullying of nature by man.
5. The presence of synthetic chemicals in, say, the human body is not itself enough to cause concern: account must be taken of the concentrations present, and the nature of the dose-response curve. Our bodies also contain traces of naturally occurring poisons, at concentrations below those associated with harm, and there is no difference in principle between synthetic and naturally occurring chemicals.
6. Given the lack of data, it is sensible to apply the precautionary principle, in all its glory – that is, including the caveat that measures taken should be cost-effective. The proposed REACH system is probably cost-effective.
7. It is crucial that regulatory regimes designed to control risk do not do so at the expense of innovation. Barriers to innovation will lead to the industry gradually moving to other parts of the world. And constant innovation is needed if we are to maintain the fight against constantly evolving microbes.
8. There are considerable complexities inherent in managing the risks from industrial chemicals in a world of traded goods. They can best be resolved by international agreement on the risk control regime.
9. Although voluntary action by industry is important in improving health and safety performance, it is most successful when accompanied by – some would say driven by – regulation, especially for product safety. The drafters of such regulation must however take full account of industry's views on the details of the proposed instruments, to ensure maximum workability and as little cost, and hindrance to innovation, as possible.
10. The proposed REACH system looks set to achieve a good level of knowledge about and management of the currently known risks from industrial chemicals (though new scares, for which chemicals have not been tested, will probably surface from time to time). The big outstanding question is whether it will turn out to have led to a decline of the European chemical sector, to the benefit of the USA and especially India and China – or whether those regions will rapidly adopt legislation with similar aims and which proves to be hardly less bureaucratic and costly than Europe's. My money is on the former.

References

1. Royal Commission on Environmental Pollution, 24th Report, Chemicals in Products, 2003, Cm 5827.

2. Speech given by Food Standards Agency Chair Sir John Krebs to the Cheltenham Science Festival, 4 June 2003.
3. T.L. Litovitz, B.F. Schmitz and K.M. Bailey, *Am. J. Emerg. Med.* 1990, **8**, 394.
4. WWF's response to the Community strategy for endocrine disruptors (undated).
5. B. Lomborg, *The Skeptical Environmentalist*, Cambridge University Press, Cambridge, 2001, Chapter 22.
6. European Environment Agency, Environment in the European Union at the Turn of the Century, 1999, http://reports.eea.eu.int/92-9157-202-0/en/page310.html.
7. R. Carson, *Silent Spring*, Houghton Mifflin, Boston, MA, 1962.
8. W.J. Hunter, B.A. Henman, D.M. Bartlett and I.P. Le Geyt, *Am. J. Ind. Med.*, 1993, **23**, 615.
9. European Commission, White Paper – Strategy for a future Chemicals Policy, COM/2001/0088 final,2001.
10. S.C. Morris and T.H. Lee, *Food Technol. Aus.* 1984, **36**, 118.
11. W. Kramer, M. Nasterlack and A. Zober, *Chemi. World,* January 2005, **46**.
12. S. Pinker, *The Blank State*, BCA, London, 2002, Chapter 13, p. 219 et seq.
13. http://www.icca-chem.org/section02a.html.

Future Perspectives in Risk Assessment of Chemicals

PETER FLOYD

1 Introduction

1.1 Overview

Some chemicals are hazardous to people and/or to the environment. The associated risks can be assessed with varying degrees of uncertainty. Within the EU, there are established procedures for assessing the risks associated with 'normal' use – and this is the focus of this chapter.

In subsequent sections, further consideration is given to the difficulties encountered in current risk assessments and how these may be addressed in the future. Of course, in the near future, it is intended that some of these difficulties will be impacted by the new regulatory regime to be established under registration, evaluation, authorisation and restriction of chemicals (REACH).

1.2 Hazard, Risk and Risk Assessment

There are many definitions of the related terms 'hazard' and 'risk' to be found in the risk literature. Simply stated, hazard is the potential for harm and risk is the probability (or likelihood) that the hazard is realized. For the purpose of this chapter, I have used the terminology adopted by the European Commission's Directorate General for Health and Consumer Protection (DG SANCO) in a comprehensive report[1] on these issues, viz:

Hazard – the potential of a risk source to cause an adverse effect(s)/event(s)

Risk – the probability and severity of an adverse effect/event occurring to man or the environment following exposure, under defined conditions, to a risk source(s).

By way of example, naphthalene presents a hazard to the aquatic compartment as it is toxic to a wide range of fish and invertebrate species. It was determined that there was a significant risk associated with the use of naphthalene in the manufacture of grinding wheels to the local aquatic environment, since the predicted environmental concentrations were significantly higher than the threshold for toxic effects.[2]

Risk assessment is a widely used methodology to provide a base to determine the level of risk associated with a particular hazard. The following DG SANCO definition (from the reference given above) will be used:

Risk assessment – a process of evaluation including the identification of the attendant uncertainties, of the likelihood and severity of an adverse effect(s)/event(s) occurring to man or the environment following exposure under defined conditions to a risk source(s).

As can be seen, risk assessment involves analysis of the hazard and derivation of the associated risk. The DG SANCO report further defines a risk assessment as comprising hazard identification, hazard characterization, exposure assessment and risk characterization, and this broad approach is followed in this paper. These various terms are defined in Table 1.

Although the underlying methodology is essentially the same across all risk assessments, the emphasis of the assessment will vary depending on the precise nature of risk under study. For example, risk assessment of an explosive facility will tend to focus on the routes by which an explosion could occur and their associated probabilities – since the consequences of an explosion of a particular size are relatively well-known. As such, the focus would be on *hazard identification* (*i.e.* the accident sequences which would lead to an explosion) and *risk characterization* (*i.e.* the probabilities that such explosions would occur). In contrast, an assessment for radon in the home would tend to focus on the degree of exposure (*i.e.* the *exposure*

Table 1 *Risk Assessment Framework (as adopted by DG SANCO)*

Stage	Definition
Hazard identification	The identification of a risk source(s) capable of causing adverse effect(s)/event(s) to humans or the environment, together with a qualitative description of the nature of these effect(s)/event(s)
Hazard characterization	The quantitative or semi-quantitative evaluation of the nature of the adverse health effects to humans and/or the environment following exposure to a risk source(s). This must, where possible, include a dose response assessment
Exposure assessment	The quantitative or semi-quantitative evaluation of the likely exposure of man and/or the environment to risk sources from one or more media
Risk characterization	The quantitative or semi-quantitative estimate, including attendant uncertainties, of the probability of occurrence and severity of adverse effect(s)/event(s) in a given population under defined exposure conditions based on hazard identification, hazard characterization and exposure assessment

assessment), since both the hazard and the effects can be readily evaluated (using established dose–response relationships for the effects of different levels of radiation). In relation to chemicals, efforts tend to be focused on establishing the degree of exposure (*i.e.* the *exposure assessment*) and a reliable dose–response relationship (*i.e.* the *hazard characterization*).

1.3 Uncertainty

Since all risk assessments are predictive, they are inherently uncertain. The attendant uncertainties (as referred to in *risk characterization*) may range from the mundane to the frontiers of human knowledge. At the simplest level, the lack of data on the existing situation relating to the usage of a particular chemical will make any estimate of the likely exposure uncertain. At the other end of the scale, despite extensive and prolonged research, the precise relationships between low levels of exposure to various chemicals and the resultant health effects, remain uncertain.

These and other types of uncertainties may be grouped as follows:

- *Knowledge uncertainty*. If we are ignorant of the mechanisms or interactions between different system components, this represents knowledge uncertainty (we do not know what we do not know).
- *Real world uncertainty*. The world we live in is characterized by uncertainty. Although the range of 'natural' conditions can be probabilistically predicted, there is always uncertainty. For example, although we can predict that 50% of our laboratory animals will succumb to an LD50 dose, we cannot predict which ones will succumb.
- *Data uncertainty*. This arises when knowledge is based on limited or incomplete sets of data, or data that may be subject to random errors. This type of uncertainty is usually expressed in terms of confidence limits (we do not have full information)
- *Modelling uncertainty*. This is determined by the validity of the methods used to predict, often in mathematical terms, possible future outcomes. These uncertainties can arise from a lack of knowledge, from the decisions made by analysts during the modelling process and from assumptions inherent within different models (models represent our best judgement).

Clearly, in dealing with a particular issue, uncertainties may arise from one type of uncertainty or a combination of the different types of uncertainties. Furthermore, categorizing uncertainty is a subjective process and there is not always a 'right' categorization. In other words, assigning a category to a particular uncertainty is not as important as recognizing that uncertainty is present.

1.4 Risk Evaluation

Risk-based decision-making requires decisions to be made as to whether or not the estimated risks associated with a particular chemical require some form of risk management. In other words, the risks (as derived from a risk assessment) need to be judged as to their potential acceptability or unacceptability. In general, the consensus is that there are three levels of risk:

- a level of risk which is so high as to demand immediate action; this is often referred to as an 'unacceptable' or 'intolerable' or '*de manifestis*' risk,
- a level of risk which is so low as to be regarded as trivial, with this being referred to as an 'acceptable' or 'negligible' or '*de minimis*' risk,
- a level of risk between these extremes, where consideration should be given to the costs and benefits of risk reduction measures.

Of course, where a risk is identified as potentially unacceptable, this should not act as the sole driver to decision-making. Other drivers, which need to be accounted for, might include sustainability, economic growth, equity and fairness, regulatory implications and political acceptability.

Although numerical risk criteria have been utilised for risks to human health across various sectors (including radiological protection and major chemical plant accidents[3]), establishing an 'acceptable' level of risks to the environment is still in its infancy.

1.5 Types of Risks

Although risks are defined in terms of probability and severity of effect, there are a range of characteristics, which can influence their acceptability or otherwise as outlined below:

- *Nature of hazard* – hazards, which are man-made, are generally regarded as being of more concern than those that are 'natural'.
- *Nature of effect* – certain effects, with particular reference to cancers, are often perceived with dread (hence the term 'dread risks').
- *Acute vs chronic* – in parallel with the above, acute (short-term) effects as a result of acute exposure are often of less concern than chronic (long-term) or delayed effects as a result of chronic exposure.
- *Reversible vs irreversible effects* – similarly, irreversible effects are generally of more concern than reversible effects.
- *Risks and benefits* – for those taking the risks also receive the benefits (for example, workers), the tolerability of risks is higher than for those who take the risks without the benefits.
- *Known vs unknown* – as would be expected, people are more concerned about risks that are difficult to understand or are very uncertain than where the risks are clearly understood.

In relation to chemicals in everyday usage, many of the associated risks possess the less favourable characteristics in that they tend to be regarded as man-made producing irreversible effects after prolonged exposure. Some are carcinogenic, mutagenic or toxic to reproduction (referred to as CMR chemicals), and the associated risks are considered to be 'dread risks'. Although modern day products offer many benefits, there tends to be an underlying assumption by consumers that the products are 'safe' and thus the associated risks of chemical constituents of these products, when identified, are often a matter of public concern. Such matters become confounded when there appears to be uncertainty as to the nature and extent of the risks. This is illustrated by the long-running debate over the safety of the use of phthalates as a plasticizer in flexible PVC products, including, for example, its use in products designed for use by babies and infants.

2 Difficulties in Risk Assessment

2.1 Overview

There are risks associated with particular chemicals in everyday usage (the risk source), and these are now subject to European risk assessment requirements. Specifically, new substances are covered by Council Directive 67/548/EEC on the Classification, Packaging and Labelling of Dangerous Substances (and amendments) and the associated risk assessment requirements are set down in Commission Directive 93/67/EEC.[4] Existing substances are covered by the Council Regulation (EEC) 793/93 on the Control and Evaluation of the Risks of Existing Substances (usually referred to as ESR) and the associated risk assessment requirements are set down in Commission Regulation (EC) 1488/94.[5]

This legislation requires that risks to both human health and the environment associated with new and existing chemicals can be considered. This is also taken to include humans exposed indirectly via the environment.

It is a standard practice to undertake a risk assessment where an adverse effect has been identified. Of course, where no risk assessment has been undertaken, this does not necessarily imply that there are no risks. One of the overarching difficulties of risk assessments is that they can take a long time to complete. As such, it is often the case that no risk assessment has been undertaken due to the resources necessary to undertake a comprehensive assessment. By way of example, since 1993, 127 ESR risk assessments have reached a first draft stage of which only 70 have been concluded![6]

2.2 Hazard Identification

As defined in Table 1, hazard identification involves the identification of a risk source(s) capable of causing adverse effect(s)/event(s) to humans or the environment, together with a qualitative description of the nature of these effect(s)/event(s).

The process of hazard identification involves the determination of what the specific adverse effects of a particular substance are likely to be. Data considered include chemical and physical properties, epidemiological studies, studies with experimental animals, *in vitro* tests and structure–activity relationships. Historically, comprehensive data have been in short supply leading to data uncertainty. By way of example, as of 1998, only 14% of the 2465 EU High Production Volume (HPV). Chemicals (*i.e.* those being placed on the EU market in quantities greater than 1000 tonnes per year per producer/importer) were considered to have a full set of relevant data[7] with most data gaps for chronic toxicity to fish and aquatic invertebrates.

Where the potential for adverse effects are unknown, this represents a classic case of knowledge uncertainty (we do not know what we do not know). By way of example, selecting endocrine disruption as a key adverse effect has only emerged during the past decade. Other effects, which at present are unknown, could include synergistic effects of two or more chemicals in combination.

2.3 Hazard Characterization

The risks must be considered for the usual anticipated conditions under which a chemical is manufactured, used and disposed of. The approach recommended by the European Commission technical guidance document[8] (TGD) for risk assessment of new and existing substances is given consideration here. It does not encompass accidental releases of chemicals or, indeed, their misuse.

Although, as defined in Table 1, hazard characterization essentially involves the development of a dose–response assessment, the focus within the TGD risk assessment methodology is on developing a threshold value below which no (adverse) effects will be observed.

In terms of human health effects, the objective is to define a 'safe' exposure level. Where the concern is over the bodily intake of chemicals, this is interpreted as a 'tolerable daily intake' (TDI), which equates to the intake (every day over a 70 year lifetime!) that will produce no adverse health effects. The ratio of the TDI to the actual/estimated intake is referred to as the 'margin of safety' (MOS).

The hazard characterization is generally accomplished by reviewing toxicological data in order to determine the highest level at which no adverse effect occurs. An assessment factor is frequently used in order to take into account inter-species toxicity differences and a final value of the 'no observed adverse effect level' (NOAEL) is obtained for the following effects:

- repeated dose toxicity and reproductive toxicity,
- acute toxicity, corrosivity and irritation,
- mutagenicity and carcinogenicity,
- skin sensitization and respiratory sensitization.

For the dose–response assessment of environmental effects, the intent is again to determine the level of exposure (*e.g.* concentration in various environmental media) at which no adverse effects would be expected to occur. The final intention is to determine a predicted no-effect concentration (PNEC) for the various environmental compartments. In the TGD and the associated *EUSES2* model,[9] the compartments are freshwater, marine, freshwater sediment, marine sediment, terrestrial, secondary poisoning (of birds and mammals) and sewage treatment plants (microorganisms).

As a broad generalization, the need for risk reduction measures is dependent on the PNEC selected. In practice, the PNEC is based on reviewing sets of experimental data that identify LOAEL (lowest observable adverse effect level) and NOEC (no observed effect concentration) values and dividing the lowest borderline value (*i.e.* below the lowest reliable LOAEL) by an 'assessment factor', which ranges from 10 to 1000 depending on the number and quality of studies available.

In practice, this often proves difficult to achieve for two principal reasons. First, levels of no-effect should be applied to complex ecosystems – which is problematic where toxicity data relate to only a small number of species. Second, toxicity data generally relate to acute effects of substances whereas chronic effects must also be taken into account in the assessment. Both of these factors have tended to require that large 'assessment factors' are applied (being greater where there is greater uncertainty as to the dose–response relationship).

The derivation of these levels is fraught with uncertainties. To some extent, numerical factors are incorporated into the assessment to account for 'knowledge uncertainty'. There are also data and modelling uncertainties which are often the focus of debates over the validity of results that are significantly lower than previous results – particularly, if the results cannot be readily reproduced.

Within the overall risk assessment methodology, it is perhaps unsurprising that this area receives the greatest degree of scientific focus. This is illustrated by the extensive work undertaken by the European Commission's Scientific Committee on Toxicity, Ecotoxicity and the Environment (CSTEE now superseded by the Scientific Committee on Health and Environmental Risks (SCHER)) in reviewing the ecotoxicological data presented in risk assessments. Nevertheless, there remains a degree of scepticism about whether risk management decisions should be based on uncertain information.

Although much of the uncertainty tends to be associated with 'data uncertainty', there may also be other uncertainties. By way of example, in relation to PFOS (perfluoroctane sulphonate), long-term mortality records showed a statistical relationship between bladder cancer and long-term working at the major US manufacturing facility. However, more detailed analysis[10] suggested that the causal link between PFOS exposure and bladder cancer (and other effects) could not be established with certainty. In other words, although PFOS could be responsible for the excess of bladder cancers, there were insufficient data on the precise exposures experienced by the workers (*i.e.* data uncertainty). However, it is also possible that the effects might be associated with other chemicals on the plant (knowledge uncertainty). There is, of course, nothing particularly unusual about this as similar accounts could be presented for many chemicals. However, what is unusual about PFOS is that the manufacturer voluntarily phased-out its production and use on the basis of its hazard characterization.

2.4 Exposure Assessment

As defined in Table 1, exposure assessment involves the evaluation of the likely exposure of man and/or the environment to risk sources from one or more media. In practice this involves four sequential steps:

- estimation of emissions to environmental compartments,
- determination of the fate of such emissions in the environment,
- derivation of predicted environmental concentrations (PECs),
- estimation of uptake by fauna (especially through the food chain to humans).

The human health exposure assessment must take into account direct exposure of workers and consumers as well as indirect exposure of the general public via the environment. Information taken into account should include measured data (for example, measured vapour concentrations); the quantity and form in which the substance is produced and used; its use pattern and degree of containment; process data, where relevant; its physicochemical properties; its breakdown and/or transformation products; likely routes of exposure and potential for absorption; frequency and duration of exposure; and the type and size of specific exposed population(s).

The environmental exposure assessment involves the determination of a PEC. This is calculated by taking into account both measured and modelled data. Usually, measured data should be used to determine the PEC value although, where this is thought to be incomplete, it should be replaced by or supplemented with results of modelling. These models are used to predict the environmental fate of a substance based upon the quantities used, releases of the substance during the various phases in its life cycle and its physical properties.

In many risk assessments, insufficient focus is placed on the emissions. In particular, while many chemical producers and formulators can produce reasonably reliable estimates (perhaps based on Emission Scenario Documents presented in the TGD or as published by OECD[11]) and/or monitoring results for emissions, the precise usage (and eventual fate) of many of the resultant products is unknown. By way of example, a variety of chemicals are used as additives in plastics, which in turn are used in numerous products. Furthermore, many such products are imported into the EU and there is the underlying concern that although, for example, certain additives should not be present in toys, such additives may be present in some imported toys.

Although some chemicals behave in a predictable manner, the behaviour of other chemicals in the environment is complex and not (fully) understood. Such knowledge uncertainty can apply to relatively common chemicals such as mercury. In the environment, mercury may be elemental, inorganic or organic but, despite extensive research, the full dynamics of the environmental behaviour of mercury are yet to be fully established.[12]

Based on the emissions and properties of a chemical, it is possible to predict the partitioning of a chemical into the different environmental compartments using the TGD or associated *EUSES2* model. This then enables the PECs (at various spatial levels) to be estimated. However, as with all risk assessment work, it is important to undertake 'reality checks' in the course of the assessment. Thus, the PECs must be consistent with the results of environmental monitoring, if reliance is to be placed upon the results.

Of course, there are examples where there are uncertainties surrounding inconsistencies between PECs and measurements. By way of example, unexpected levels of decabromodiphenyl ether (DecaBDE) were found in a collection of birds' eggs,[13] which potentially were a cause for concern. Three explanations can be advanced for such findings:

- birds' exposure to DecaBDE was much greater than expected,
- unknown sources of DecaBDE encountered during storage led to increased levels in birds' eggs,
- environmental behaviour of DecaBDE is not fully understood.

Such possibilities can be eliminated through further study – for example, checking levels of DecaBDE in other egg collections would assist in determining whether the findings were collection-specific or not.

For both human health and environmental exposure assessments, the calculations are facilitated by the provision of default values. There are over 200 default values for *EUSES2* covering numerous parameters including the organic carbon content of soil (2%); the area fraction of freshwater (3%); the average wind speed (3 m s^{-1}); and drinking water intake for cattle (55 l day^{-1}). In relation to human dietary intake, such default assumptions are carefully considered as these have a direct effect on the resultant risks. By way of example, the Scientific Committee on Food has declared[14] that a daily intake of 200 g of fat (in food) can be considered to be "a realistic maximum". This is important in determining the (maximum) human uptake of various organic chemicals, which disperse through the environment and enter the food chain.

2.5 Risk Characterization

As defined in Table 1, risk characterization normally results in the quantitative or semi-quantitative estimate, including attendant uncertainties, of the probability of occurrence and severity of adverse effect(s)/event(s) in a given population under defined exposure conditions.

This involves an estimation of the incidence and severity of the actual or potential effects, including quantification of the likelihood of effects where appropriate. Within the context of the TGD, this process differs slightly between the human health and environmental risk assessments.

In terms of human health, the risk characterization is accomplished by comparing the NOAEL/TDI values for the various effects with the dose or concentration to which people (workers and consumers) are exposed. It is common practice to express the relationship between the two in terms of MOS, which is the ratio of the predicted exposure to the NOAEL/TDI. The magnitude of, and certainty in, the MOS is used to form a basis for risk management.

Environmental risks are characterized by comparing the ratio of the PEC/PNEC for each of the compartments of interest. Where this value is greater than one, a basis can be formed for action to reduce the risks, dependent upon the degree of certainty in the PEC and PNEC values.

In addition, risks to the general public posed through exposure via the environment are also required to be assessed. This involves assessing the concentrations of a substance in various intake media (air, food, etc.), the rate of intake of each medium and a combination of the two in order to assess the exposure (including a factor for bioavailability, where appropriate). The exposure is then compared to the NOAEL/TDI used in the standard human health risk assessment process. This is essentially a combination of the human health and environmental approaches.

Under ESR chemical risk assessments, the risk characterization (for both human health and the environment) is presented in the form of one of three possible conclusions:

- *Conclusion (i).* There is need for further information and/or testing.
- *Conclusion (ii).* There is at present no need for further information and/or testing and no need for risk-reduction measures beyond those, which are being applied already.

- *Conclusion (iii).* There is a need for limiting the risks; risk-reduction measures, which are already being applied shall be taken into account.

Broadly speaking, these conclusions represent levels of risk, which are potentially significant, negligible and significant, respectively. Thus, where there is a potentially significant risk to either human health or the environment (in other words, Conclusion (i)) then action should be taken to reduce uncertainty in order to reach a more robust conclusion (*i.e.* Conclusion (ii) or Conclusion (iii)). Where a significant risk (*i.e.* Conclusion (iii)), then action should be taken to mitigate that risk – generally through restrictions on use. However, there are concerns (particularly from industry) that the methodology employs overly conservative default assumptions, which lead to overestimates of the risks and, hence, unnecessary risk mitigation actions. Conversely, campaigning groups (such as WWF, Greenpeace and Friends of the Earth) argue that particularly hazardous chemicals found in the environment should be restricted in any event.

However, prior to imposing such actions, the European Commission undertakes a risk-benefit analysis of the potential implications. In other words, will the reduction in risks achieved by the proposed actions outweigh the costs and risks associated with introducing new processes and/or substitute chemicals?

In summary, there are few perceived difficulties with the risk characterization stage itself. Rather the debate is over the degree of uncertainty associated with both the no effect levels and the predicted concentrations and uptakes.

3 Current Developments

3.1 Overview

Across the EU, various approaches to aspects of the risk management of chemicals in everyday usage are being developed. Following the 2001 White Paper on a future chemicals policy,[15] the European Commission adopted the proposal for the REACH in October 2003.[16] Although the proposal is still subject to ongoing negotiations, there is little doubt that the implementation of REACH will be the most important development in this field in the foreseeable future and will produce significant advances in data availability and consistency.

3.2 Hazard Identification

As indicated in Section 2.2, there are significant data gaps on the hazards associated with chemicals in everyday usage and this issue will be addressed, in part, by REACH.

A basic requirement of REACH is that everyone of the estimated 30,000 substances placed on the market that are manufactured in (or imported into) the EU in quantities, in excess of 1 tonne per year, will need to be registered. Each registration will need to be accompanied by a Technical Dossier,[17] which in time for substances manufactured/imported in quantities of more than 10 tonnes per year, will incorporate a chemical safety report (CSR), which will detail a chemical safety assessment (CSA). The requirements will be introduced progressively with initial priority given to assessments of CMR and high-volume (greater than 1000 tonnes per year) substances.

It is acknowledged by the Commission[18] that this approach is pragmatic and that some lower-volume substances, which are particularly hazardous, may not be assessed in the early stages of REACH implementation. However, REACH proposes that 'substances of very high concern' will need to be authorized (irrespective of quantities) and such authorization will, *inter alia*, be subject to a CSR. It is proposed that these substances will include CMRs, persistent bioaccumulative and toxic (PBT)/very persistent and very bioaccumulative (vPvB) (discussed further in Section 3.3) and other substances of equivalent concern.

Candidate substances for authorization include endocrine disruptors, which are currently being extensively researched in many countries. Indeed, within Europe, this has become a major multi-million pound industry with the establishment of the £15 m (€20 m) Credo research project across 64 research teams in 2002.[19]

Of the factors outlined earlier, which influence the acceptability of the risks, endocrine disruptors have all the 'worst' characteristics since they are associated with man-made materials and they produce 'dread' irreversible effects after chronic exposure. Furthermore, not only are the effects highly uncertain and difficult to understand, but they occur among the unborn and newly born (in other words, it would appear that the few, if any, of the benefits of exposure are likely to be outweighed by the risks). By way of example, recent results that endocrine disruption (associated with exposure to phthalates) might be being observed in humans provided media headlines of the type: *Toxin in plastic harming unborn boys – scientists say chemicals have gender bending effects.*[20]

As with other 'topical' subjects such as climate change and genetically modified organisms, cynics might argue that while the 'research industry' should be impartial, it does have a vested interest in the outcomes of such research – as, of course, do large corporations, governments, etc.

The procedures by which hazards should be identified and prioritized under REACH are currently the subject of various RIPs (REACH implementation projects) being co-ordinated by the European Chemicals Bureau.

3.3 Hazard Characterization

It is intended that the implementation of the REACH proposals will lead to a more comprehensive data set on the chemicals used in today's Europe. This will be a welcome change for those attempting to undertake risk assessment work using IUCLID (International Uniform Chemical Information Database) data sheets with numerous gaps on, seemingly, the most basic physico-chemical properties. In essence, where there are important data gaps, the manufacturer/importer may be obliged to undertake (further) testing to obtain the necessary information with associated administrative, technical and financial implications.

There are, of course, a variety of 'screening' aids, which have been developed to prioritize those substances that are considered particularly hazardous. In the last decade, considerable attention has been focused on PBT and/or vPvB substances using the characteristics shown in Table 2.

Currently, 74 substances are under review by the EU-15 Member States (and Norway) with each country acting as a rapporteur for several PBT/vPvB candidate substances. It is important to stress that the PBT/vPvB classification was developed

Table 2 *PBT and vPvB criteria (from the TGD)*

Criterion	PBT Criteria	vPvB Criteria
P (persistent)	Half-life > 60 d in marine water or > 40 d in freshwater* or > 180 d in marine sediment or > 120 d in freshwater sediment*	Half-life > 60 d in marine or freshwater or > 180 d in marine or freshwater sediment
B (bioaccumulative)	BCF > 2000	BCF > 5000
T (toxic)	Chronic NOEC < 0.01 mg/l or CMR or endocrine disrupting effects	Not applicable

*For the purpose of marine environmental risk assessment half-life data in freshwater and freshwater sediment can be overruled by data obtained under marine conditions.

as a means to protect the marine environment from particularly harmful substances (as set out in the TGD). This can create confusion when collating data for the associated *EUSES2* model, which is primarily focused on the freshwater compartment. In other words, data collated for *EUSES2* may differ from that used in a PBT screening assessment. This is exemplified by organotins (notably tri-butyl tin) where the values of P, B and T differ significantly between the marine and freshwater environments.

There are cases where the use of PBT/vPvB criteria can be problematic. Since metals are, of course, inherently 'very persistent' (notably in sediments), the use of the PBT/vPvB criteria presents particular problems for these substances. Industry has argued that the issue is further complicated by the very low solubility of most metals, which in turn leads to low bioavailability in the aquatic environment.[21] Furthermore, the relationship between the BCF (bioconcentration factor) and exposure concentrations for metals is quite different from that for organics, which it is claimed, makes the BCF factor an unreliable indicator of the hazard potential of metals.[22]

It is interesting to note that the PBT/vPvB criteria proposed in Annex XII of the REACH proposal differ slightly from those presented in Table 2. The REACH criteria are shown in Table 3 with the differences highlighted in ***bold italics***.

Confusingly, the REACH proposal for including 'substances of very high concern' into Annex XIII (list of substances subject to authorization) as set out in Article 54 is that the substances should meet one (or more) of the following criteria:

- substances are classified as CMR,
- substances are PBT or vPvB,
- *'substances, such as those having endocrine disrupting properties ... identified as causing serious and irreversible effects to humans or the environment which are equivalent to those of other substances listed* [above] ...' (Article 54(f)).

It is likely that such wording will lead to intensive debates (and lobbying) as to which groups of substances should be included/excluded from Annex XIII. Within the UK, formal proposals are already under discussion as to whether the 'equivalence' lists should include substances, which are vvB, vBvT, vvT, endocrine disruptors or found at significant levels in the environment.[23]

Table 3 *Proposed PBT and vPvB criteria (from Annex XII, REACH proposal)*

Criterion	PBT Criteria	vPvB Criteria
P (persistent)	Half-life > 60 d in marine water or > 40 d in fresh/*estuarine* water or > 180 d in marine sediment or > 120 d in fresh/*estuarine* water sediment *or > 120 d in soil*	Half-life > 60 d in marine or fresh/*estuarine* water or > 180 d in marine or fresh/*estuarine* water sediment or *> 180 d in soil*
B (bioaccumulative)	BCF > 2000 *(measured value)*	BCF > 5000
T (toxic)	Chronic NOEC < 0.01 mg/l or CMR or *there is other evidence of chronic toxicity, as identified by the classifications: T, R48, or Xn, R48 according to Directive 67/548/EEC*	Not applicable

Note: For completeness, the classification T, R48 is defined as: Toxic – danger of serious damage to health by prolonged exposure; and Xn, R48 as: Harmful – danger of serious damage to health by prolonged exposure.

Whichever substances are prioritized, there will be a need for additional data to provide sufficient information for robust characterizations of the hazards and the associated risks. One means to generate such data is the use of quantitative structure-activity relationships (QSARs), which have been developed over the last two decades. In essence, these allow parameters to be predicted for a particular chemical based on an analysis of the results for a range of other chemicals. OECD maintains a comprehensive database of QSAR models[24] and other sources (with data sets) include the European Chemicals Bureau.[25] Within the EU, the most comprehensive trial of QSARs was undertaken by the Danish EPA, which concluded that the reliability of the models used was greater than 70% after trials on 47,000 substances.[26]

There are, however, a number of difficulties with QSARs. For a particular chemical, it is possible to predict a range of parameters (including, for example, physico-chemical properties and toxicity effects) with varying degrees of (modelling) uncertainty. However, in some cases, there may be a poor correlation between the 'traditional' QSAR predictions and the validation measurements for a broad group of chemicals. In a recent study on 400 pesticides,[27] poor correlations were observed not only for toxicity but also for basic parameters such as melting point, boiling point and vapour pressure. On the other hand, reasonable correlations were found for solubility and partition coefficients. While arguments could be advanced that pesticides are a 'special' case, the current uncertainties in QSAR predictions ensure that the predictions cannot be relied upon for the purposes of regulation. However, it is interesting to note that the cost estimates for REACH are based on the assumption that validated QSARs, will become available to provide some of the data requirements.[28]

3.4 Exposure Assessment

One of the main impacts of REACH will be to develop a more comprehensive picture of the usage of hazardous substances in everyday products through the involvement of 'downstream users'. Specifically, the manufacturer/importer will be required to develop 'exposure scenarios' (which encompass emissions, pathways and exposure levels) based on the known uses. Where downstream users are aware of limitations of such exposure scenarios, there are essentially two ways forward:

- either, the downstream user performs its own CSA,
- or, improved information is communicated 'up' the supply chain to the manufacturer/importer to provide the necessary information to revise its CSA.

While such requirements should significantly enhance the quality of exposure assessments, there is the potential for the refinement of the manufacturer/importer's CSA never to be concluded as there could be ever-changing use patterns in the market place (across all 25 EU Member States). However, even if the uses can be fully described, there will still be numerous areas of uncertainty. To take a simple example, if one downstream user estimates emissions to be 1% and another estimates 0.1% for the same new use – how should the CSA be refined?

Even in the ideal world where there is consensus over emissions, environmental fate and the resultant PECs, estimating the uptake by humans and other organisms can be problematic. By way of example, Greenpeace has published various reports on levels of chemicals in buildings and products.[29,30] While the presence of various chemicals in, say, house dust provides a useful indicator of how chemicals are spread through the environment, another set of assumptions and calculations are required to use such information to estimate their likely human uptake.

3.5 Risk Characterization

For many chemicals, the focus is on human health risks. In some cases, it is relatively straightforward to define the risks to the 'average' consumer as well as the risks to a defined sub-group of 'high-risk' consumers. This high-risk group may be at risk because of their dietary intake or because of their age. For example, consumer exposure to tri-butyl tin is dominated by its presence in seafood and the risk can be characterized for those with a high seafood intake — and this can be defined as the 95th percentile value.[31] The situation becomes more complex when dealing with numerous exposure pathways. Common plastics such as PVC can contain a range of additives and are used in a wide variety of products. In these cases, although it is possible to build up a picture of the 'worst case' consumer who uses all the identified products, it is, of course, an overly cautious and unrepresentative scenario to be used as basis for risk-based decision-making.

For risks to the environment, the risk characterization stage simply involves dividing one value (the PEC) by another (the PNEC) for each compartment of interest. In most cases, and as would be expected, debate focuses on the robustness of these values and the associated assumptions. One area where there is debate over the approach to risk characterization is for substances with a 'natural' background such as metals. If the background concentration is B, then the PEC value will be B plus environmental concentrations resulting from the uses under study (PEC*add*). Similarly, there will be a permitted additional concentration (PNEC*add*) so that the

PNEC = B + PNEC*add*. As might be expected, although the principle is sound, defining a 'background' value across the EU for each environmental compartment is problematic. Of course, matters are made worse when B exceeds the PNEC as this implies that the natural background level already presents a significant risk, as is the case for nickel.[32]

4 Future Perspectives

4.1 Overview

There will continue to be requirements for the assessment of the hazards and risks associated with chemicals in everyday usage into the foreseeable future. In the short-to-medium term, there will be significant advances in data availability as a result of REACH, which it turn are likely to lead to refinements of the predictive models used in risk assessment. In the longer term, it is likely that new hazards and risks will emerge and new methodologies will be developed to assess them.

4.2 Hazard Identification

In simple terms, the spotlight of concern has progressively focused on chemicals, which are:

- acutely toxic to humans (for example, chlorine and ammonia),
- chronically toxic to humans (for example, asbestos and lead),
- toxic to humans at very low concentrations (for example, dioxins and furans),
- toxic to the environment (chemicals in general),
- responsible for developmental effects (endocrine disrupters).

To date, risk assessments have tended to concentrate on the risks associated with single chemicals. It is entirely possible that consideration of chemicals in combination could lead to a more probabilistic approach – for example, what is the probability that chemical X will be encountered in combination with chemical Y to produce synergistic/antagonistic effects? Although some work has been done in this field, it is far from sufficiently developed to be integrated into regulation.[33]

One area for a new generation of hazards is likely to be the use of various chemicals in nanotechnology. Already, there are concerns being expressed that nanoparticles and nanotubes can cross the human body 'barriers' and enter the brain and other critical organs. Although there are suggestions that such nanomaterials could be allocated CAS (Chemical Abstracts Service) numbers and treated as 'new chemicals' under REACH,[34,35] it would also be possible to develop a separate regulatory regime with the requirements tailored to the potential hazards and risks associated with nanotechnologies — as was done for genetically modified materials.

4.3 Hazard Characterization

In the short term, the implementation of REACH will lead to the generation (and publication) of comprehensive data on a few thousand high-volume substances. It is to be hoped that this much larger data set will enable much more reliable QSARs to be developed, thus, reducing the (potential) need for extensive testing of lower volume substances in the medium term. In parallel with the development of QSARs, it is likely that concerns over extensive animal testing will also lead to further development of alternative testing methods. A register of currently validated methods is maintained within the EU by the European Centre for the Validation of Alternative Methods (ECVAM).[36]

It is likely that this additional data generated under REACH will also facilitate discussions on P, B and T characteristics. However, it must be remembered that the use of PBT/vPvB criteria should be regarded as a means to screen and prioritize those chemicals, which merit further study. In other words, the use of these criteria (which are essentially a form of hazard rating) for regulation would not constitute risk-based decision-making (since no account is taken of exposure).

In the longer term, it is likely that there will be moves by international bodies to standardize the criteria to be used for the nature, quality and utility of relevant scientific evidence. One example of this is the recent call by FAO/WHO (Food and Agricultural Organisation/ World Health Organisation) to establish guidelines for the risk assessment of nutrients.[37] As has already been noted (in Section 2.4), OECD is involved with the production of internationally accepted emission scenario documents but it is also involved with the development of harmonized testing procedures, QSARs, classification and labelling of chemicals and risk management guidance documents. It is of note that OECD is working towards a global portal for hazard data on HPV chemicals.[38]

4.4 Exposure Assessment

As already indicated, the exposure assessment involves four sequential steps:

- estimation of emissions to environmental compartments,
- determination of the fate of such emissions in the environment,
- derivation of PECs,
- estimation of uptake by fauna (especially through the food chain to humans).

It is to be hoped that REACH and other developments such as increased environmental monitoring will result in a much greater understanding of the uses, emissions and environmental fate of chemicals. However, considerable work will be required to ensure that the results of predictive models for PECs and consequential uptake by fauna are consistent with reality – which is a pre-requisite for robust risk assessments.

At a practical level, the enlargement of the EU from 15 to 25 Member States will require the default assumptions of the TGD/*EUSES2* to be revisited to ensure their validity across the EU-25. Similarly, although the wider involvement of industry in assessing the chemical risks associated with their products is to be welcomed, few would argue that the use of the TGD/*EUSES2* in its current form is accessible to the non-expert. As such, there will be a need for more user-friendly techniques to assist those who are not experts in risk assessment but, nevertheless, have a responsibility for the assessment and evaluation of the risks (and the hazards) associated with chemicals in their products. It may be the case that this will result in a generally accepted screening process (such as use of PBT criteria) to ensure that resources are focused on the risks of greatest concern to people and to the environment.

4.5 Risk Characterization

It is unlikely that there will be much change in the general approach to risk characterization since the underlying assumption that if the predicted level exceeds the no-effect level then there will be a risk is difficult to challenge. As at present, arguments will continue as to the reliability of the predicted and no-effect levels used in the risk assessment.

5 Conclusions

Within the EU, there are established procedures for undertaking risk assessments of chemicals in everyday usage. Although there is a general recognition that the risk assessments are inherently uncertain, the focus of scientific attention tends to be on the hazard characterization (*i.e.* dose – response relationships). In practice, there are also many uncertainties associated with emissions and exposure pathways (*i.e.* the exposure assessment).

There is little doubt that the implementation of REACH will reduce uncertainties associated with the use patterns of chemicals and will substantially increase the body of available effects data. Such outcomes will facilitate the development of more reliable risk assessments – particularly if the increase in data enables the development of more robust QSARs to be used for the reliable prediction of effects.

More generally, it is likely that approaches will be developed and agreed at an international level to tackle problem areas such as metals and effects at very low levels. However, it is likely that, in the longer term, 'new' hazards will be identified, which will require new data and new approaches.

Due to the variety of end points considered in an assessment of the risks associated with chemicals to people and to the environment, the risk assessment process is already very complex. Given the scale of the potential numbers of risk assessments required, there is a need for more user-friendly approaches to risk assessment and for generally accepted simplified screening procedures so that resources are focused on the most significant risks.

Finally, it must be emphasized that risk assessments are intended to assist with risk-based decision-makers. As such, any identified hazards and risks need to be balanced against the associated benefits before decisions are made.

References

1. European Commission, First Report on the Harmonisation of Risk Assessment Procedures, DG Health & Consumer Protection, 2000.
2. Health & Safety Executive/Environment Agency, Naphthalene Summary Risk Assessment Report, ESR RAR, 2003, (available from http://ecb.jrc.it).

3. D.J. Ball and P.J. Floyd, Societal Risk, Report prepared for the UK Health & Safety Executive, London, HSE Books, 1999 (also available from www.rpaltd.co.uk).
4. Commission Directive 93/67/EEC of 20 July 1993 *Laying down the Principles for the Assessment of Risks to Man and the Environment of Substances notified in accordance with Council Directive 67/548/EEC.*
5. Commission Regulation (EC) 1488/94 of 28 June 1994 *Laying down the Principles for the Assessment of Risks to Man and the Environment of Existing Substances in accordance with Council Regulation (EEC) No 793/93.*
6. *European Chemicals Bureau Newsletter*, dated 19 May 2005. The associated reports are available from the Online European Risk Assessment Tracking System (ORATS) –(http://ecb.jrc.it).
7. R. Allanou, B.G. Hansen, and Y. van der Bilt, *Public Availability of Data on EU High Production Volume Chemicals*, European Chemicals Bureau report EUR 18996 EN, 1999.
8. *European Commission, Technical Guidance Document in Support of Commission Directive 93/67/EEC on Risk Assessment for New Notified Substances and Commission Regulation (EC) No 1488/94 on Risk Assessment for Existing Substances and Directive 98/8/EC of the European Parliament and of the Council* concerning the Placing of Biocidal Products on the Market, Luxembourg, Office for Official Publications of the European Communities, 2003.
9. *EUSES2* is the computer program, *European Union System for the Evaluation of Substances,* developed by RIVM in 2003, which codifies the TGD equations.
10. OECD, Hazard Assessment of Perfluorooctane Sulfonate (PFOS) and its Salts, OECD Environment Directorate Report, dated 21 November 2002 (available from www.oecd.org).
11. OECD, *Emission Scenario Document on Plastics Additives*, dated 24 June 2004.
12. UNEP, Global Mercury Assessment Report, December 2002 (available from www.chem.unep.ch).
13. Environment Agency, Bis (Pentabromophenyl) Ether Risk Assessment Report, ESR RAR, 2003 (available from http://ecb.jrc.it).
14. Scientific Committee on Food, *Opinion on the Introduction of a Fat (Consumption) Reduction Factor (FRF) in the Estimation of the Exposure to a Migrant from Food Contact Materials*, CS/PM/3998, Final dated 12 December 2002, DG SANCO.
15. COMM (2001) 88 Final, White Paper on the Strategy for a Future Chemicals Policy, adopted by the European Commission on 13 February 2001.
16. *COM 2003 0644, Proposal for a Regulation of the European Parliament and of the Council concerning the Registration, Evaluation, Authorisation and Restriction of Chemicals (REACH), Establishing a European Chemicals Agency and Amending Directive 1999/45/EC and Regulation (EC) {on Persistent Organic Pollutants}; Proposal for a Directive of the European Parliament and of the Council Amending Council Directive 67/548/EEC in order to adapt it to Regulation (EC) of the European Parliament and of the Council concerning the Registration, Evaluation, Authorisation* and Restriction of Chemicals, adopted by the European Commission on 29 October 2003.

17. European Commission, *The REACH Proposal – Process Description*, information document dated June 2004.
18. European Commission, *Questions and Answers on REACH*, information document dated 22 November 2004.
19. European Commission, *Tracking Down Endocrine Disrupters*, news article dated 24 February 2004 (available from http://europa.eu.int/comm/research/newscentre).
20. Front page of *The Guardian*, 27 May 2005.
21. D.M.Di Toro, C.D. Kavvadas, R. Mathew, P.R. Paquin and R.P. Winfield, *The Persistence and Availability of Metals in Aquatic Environments*, International Council on Metals and the Environment, Ottawa, 2001.
22. J.C. McGeer *et al.*, Inverse Relationship between Bioconcentration Factor and Exposure Concentration for Metals: Implications for Hazard Assessment of Metals in the Aquatic Environment, Environ. Toxicol. Chem., 2003, **22**(5), 1017–1037.
23. UK Chemicals Stakeholder Forum, Substances of Equivalent Concern – A Draft Definition from the Advisory Committee on Hazardous Substances, discussion paper presented to the UK CSF, Meeting of 25 January 2005.
24. Models may be accessed via *OECD's Database on Chemical Risk Assessment Models* (available from http://webdomino.oecd.org/comnet/env/models.nsf.).
25. Access to ECB's information on QSARs and associated data sets (available from http://ecb.jrc.it/QSAR).
26. Miljøministeriet (Danish EPA), Report on the Advisory List for Self-classification of Dangerous Substances, Environmental Project No. 636, 2001 (available from www.mst.dk).
27. O. Hansen, Quantitative Structure-Activity Relationships (QSAR) and Pesticides, Miljøministeriet (Denmark) Pesticides Research Report No. 94, 2004 (available from www.mst.dk).
28. RPA, Revised Business Impact Assessment for the Consultation Document, Working Paper 4, Report prepared for DG enterprise and dated October 2003 (available from http://europa.eu.int/comm/enterprise/ reach/eia_en.htm).
29. D. Santillo, I. Labunska, H. Davidson, P. Johnston, M. Strutt and O. Knowles, *Consuming Chemicals: Hazardous Chemicals in House Dust as an Indicator of Chemical Exposure in the Home*, Greenpeace Research Laboratories, dated January 2003.
30. TNO, Hazardous Chemicals in Consumer Products, Report for Greenpeace dated September 2003.
31. F. Willemsen, J-W. Wegener, R. Morabito and F. Pannier, Sources, Consumer Exposure and Risk of Organotin Contamination in Seafood, Report for the European Commission by the Dutch Institute for Environmental Studies, dated December 2004.
32. As outlined in *Nickel in the Environment – Current Information* on the Nickel Forum (available from www.enia.org).
33. F.M. Christensen, J.H.M. de Bruijn, B.G. Hansen, S.J. Munn, B. Sokull-Klüttgen and F. Pedersen, Assessment Tools under the New European Union Chemicals Policy, Greener Manage. Int., 2003, (41), 5–19.

34. The Royal Society & The Royal Academy of Engineering, *Nanoscience and Nanotechnologies: Opportunities and Uncertainties*, The Royal Society & The Royal Academy of Engineering, London, 2004.
35. European Commission, Nanotechnologies: A Preliminary Risk Analysis, Report by DG Health & Consumer Protection based on a Workshop held in Brussels, 1–2 March 2004.
36. European Centre for the Validation of Alternative Methods (available from http://ecvam.jrc.it).
37. FAO/WHO, Background Paper and Request for Comment/Call for Information, Issued as part of the *Development of a Scientific Collaboration to Create a Framework for Risk Assessment of Nutrients and Related Substances*, dated October 2004.
38. OECD, *Environment, Health and Safety News*, OECD Chemical Safety newsletter No. 17, dated April 2005.

Assessing Risks to Human Health

PAUL T. C. HARRISON AND PHILIP HOLMES

1 Introduction

1.1 Chemicals in the Environment

Chemicals in the environment may be natural or synthetic, organic or inorganic, may arise from point or diffuse sources, and be locally or globally distributed. Human exposure can be through a multitude of routes and may be a result of deliberate release, accidental release or the everyday use of chemicals.[1] Depending on the physicochemical attributes and behaviour of the chemical and its environmental distribution, exposure can occur through inhalation of polluted air (either indoors or outdoors), direct ingestion through water or food, ingestion of material through pica or accidental transfer to food materials, or by skin contact.

Most concern about exposure to chemicals in the environment from the human health perspective is directed towards those that are man-made, although many naturally occurring substances are highly toxic and some chemicals of concern, such as dioxins, can arise from both natural processes (*e.g.* forest fires) and human activities. Modern techniques of chemical analysis now allow exceedingly low levels of chemicals to be routinely detected in the environment,[2] which makes it increasingly important to understand the concepts of exposure and dose, and the difference between hazard and risk.

1.2 Hazard and Risk

Hazard describes the inherent potential of a substance to cause harm. Thus if, for example, a chemical is highly acidic or highly alkaline, it has the *potential* to cause burns to the skin and therefore has intrinsic hazardous properties. Risk, on the other hand, is a measure of the likelihood or extent of harm associated with a given hazard, and in order to be able to characterise the risk for any given situation, information is needed on both the hazard potential and the degree of exposure. If there is no possible exposure pathway to humans, then there can be no risk to human health. Assessment of risk requires full understanding of both hazard (*e.g.* toxicity) and exposure (including pathways, exposure levels and dose).[3,4] Risk management typically follows from the assessment of risk and an appraisal of the options available for its reduction or mitigation; this process may be heavily influenced by other, non-scientific, factors such as the social and economic importance of the chemical and the public's perception of the nature of the risk.

1.3 Risk Perception

The perception of risk is important in chemical policy development and is driven by a number of factors. Whereas the professional risk analyst will evaluate the degree of risk posed by a hazard in terms of the results of a risk assessment and consideration of the risk management options, the majority of the public relies on intuitive judgements.[3] Thus, different people can be expected to perceive risk differently, depending on the likelihood of adverse effects, whom it affects, how familiar or widespread the effect is, and whether the individuals have voluntarily agreed to bear the risk and if they perceive potential benefits to themselves.[3] Very often, risks that are unknown, involuntary and/or dreaded are perceived as greater and more important than risks that are common, voluntary and well understood.

While this chapter deals predominantly with the assessment of risk based on standard scientific principles, it is fully acknowledged that risk perception is an important element in the overall risk assessment and management paradigm that can impact on the final risk management solution selected, thus underlying the need for risk assessors to ensure that risks and the possible options to address them are fully and clearly communicated to all stakeholders.

2 Chemical Hazard Assessment

In assessing risks posed by chemicals to the environment, there are certain characteristics that are commonly taken to be important. These parameters are: persistence (P), the propensity for a substance to withstand degradation and therefore remain in the environment in an unchanged state for a prolonged period of time; bioaccumulation (B), the propensity to build up in biota (through, for example, accumulation in fatty tissues) resulting in bioconcentration that can especially impact top predators; and toxicity (T), resulting in measurable harm to organisms in the environment (normally measured in aquatic organisms). In addition, chemicals are considered to be of concern if they are carcinogenic (C), that is they cause neoplasms, or are mutagenic (M) with the ability to cause cancers or affect the heritable genetic material, or are toxic to reproduction (R), causing measurable impacts on reproductive function (*e.g.* fertility) or outcome (*i.e.* defects in offspring).

The formal assessment of hazard to humans from chemicals in the environment focuses not only on the C, M and R properties of a chemical but also includes consideration of acute and chronic toxicity through the relevant routes of exposure (oral, dermal and inhalation, as appropriate). There are various standardised laboratory tests, validated through the Organisation for Economic Cooperation and Development (OECD),[5] that are designed to address different toxic endpoints by various exposure routes. Testing for general toxicity may extend from acute (short-term, *i.e.* one or a few days) to chronic (greater than 13 weeks) while, in some cases, reproductive effects may be investigated over two generations or more. Where exposure via the air is expected to be important, the inhalation route may be used in the toxicity testing; if skin contact is likely, assessment of irritation and sensitization potential is likely to be included. An aspect of much current uncertainty as to its toxicological significance is endocrine disruption, by which a substance has the potential to interact with, or otherwise adversely affect, hormonal systems.[6] To some scientists, endocrine disruption is an important toxic endpoint in its own right, while others see it simply as a perturbation in normal physiological processes that may or may not lead to deleterious effects in the intact organism, which would be identified during routine testing for reproductive toxicity. Although examples of endocrine disruption are well established in a number of wildlife species, and there is considerable evidence for temporal changes in human reproductive parameters, such as sperm quality and quantity, testicular cancer, prostate cancer and breast cancer,[7–9] the actual impact of environmental chemical exposure on these trends remains highly contentious. The determination of an agreed validated 'suite' of screening and testing methods to ascertain whether a substance may have adverse impacts on the reproductive system or major endocrine systems (such as the adrenal, pituitary and/or thyroid) is fraught with difficulty – and has not yet been achieved.[10]

The issue of 'alternative testing', that is replacing the use of animals in the process of toxicity assessment by other (*e.g. in vitro* or computer) methods that do not use animals, is a much exercised question, and a number of organisations have published reports and commentaries on this. Certainly, there are invigorated efforts to reduce animal testing and a number of ongoing activities in Europe specifically aimed at developing alternative methodologies, but it seems likely that such efforts will take many years to bear fruit.[11]

Another issue that is often raised in considering effects of chemicals on human health is the question of multiple exposure and the effects of 'mixtures', since environmental exposure to a chemical inevitably occurs in some kind of mixture. Simultaneous exposure to more than one substance can result in simple additivity of the individual effects, or the substances may be antagonistic, with a less than additive result, or synergy may occur, whereby the effect of one substance is markedly increased by the presence of another. The way in which substances 'interact' like this is at least partly determined by the individual biological mechanisms of effect of those chemicals. Because of the current toxicity testing paradigm, it is currently difficult or impossible to address the question of mixtures experimentally (simply because of the number of possible combinations and permutations) but specific laboratory and mathematical techniques are now being developed that can at least provide a handle to begin addressing this question.

3 Assessing Risk

The assessment of risk from chemicals in the environment involves consideration of a raft of properties, principles and characteristics. The first step is normally to assess the toxic properties of the substance, based on published literature and/or test results. Establishing toxicity requires not only an understanding of the behaviour of the chemical in the body – where it is distributed and how it is metabolised, for example – but how that behaviour may differ between man and the particular test species/system used (if the data are not derived directly from human experience). It is also necessary to establish the mechanisms of action, which may be important in assessing the organ(s) affected and therefore the likely consequences in man.

Of paramount importance in the assessment of risk is the establishment of the relative significance of the various possible exposure pathways (*e.g.* gastrointestinal tract, respiratory system or skin), because this determines not only the extent to which various tissues might be exposed and thus affected,[12,13] and therefore which toxicity testing is most relevant, but also whether significant exposure is likely to occur at all. Exposure then needs to be quantified so that environmental exposure levels can be compared to the exposures used in test systems, enabling the determination of whether or not the environmental levels are likely to do harm. If toxicity data indicate a hazard, and analysis of exposure pathways and levels show that exposure levels may be such as to raise the possibility of harm (allowing for uncertainties in extrapolation – see Section 8), then careful assessment of this information will allow judgments to be made about likely effects in individuals or populations.

The science of epidemiology allows the predictions of risk assessments based on toxicity study assessments of hazard to be tested in human populations and is therefore potentially a very useful and persuasive tool, except that problems with confounders (discussed further in a later section) often make interpretation of the results difficult. Sometimes the epidemiological studies come first and identify possible associations between exposure to an environmental factor and an adverse health effect. In such cases, these associations are used to generate hypotheses concerning cause and effect linkages that can then be tested by standard risk assessment practices, based on the use of toxicological and exposure assessment principles, to establish whether the apparent associations revealed by epidemiological studies are biologically plausible.[14]

4 Qualitative vs. Quantitative Risk Assessment

As explained above, the basic underlying approach to risk assessment for chemicals involves integration of knowledge on the hazardous properties and dose–response relationship for a substance (based on human and/or experimental data), with knowledge or estimates of the level (and, if appropriate, route) of likely exposure, to characterise the likelihood of an adverse response occurring under a given set of conditions. If possible, risk characterisation should also seek to assess the uncertainties that may surround the risk estimate produced.[15] Historically, risk assessment was frequently limited to the presentation of a narrative that described the elements of the risk assessment paradigm for a particular hazard, and then attempted to express, supported by expert judgement, the degree of risk anticipated using purely qualitative terms, such as "negligible", "minimal", "moderate" or "severe". Increasingly, in order to provide a clearer and more useful understanding of the degree of risk and the uncertainty surrounding an estimate, methods that allow presentation of risk in numerical (quantitative) terms (*e.g.* % change in response per mg/kg bodyweight of exposure) are applied. However, it is more usual to find that any given risk assessment will still incorporate some element of qualitative judgement as well as quantitative elements.[16]

Many methods of quantitative risk assessment have been developed, with different regulators in various countries adopting different approaches and methodologies. Although detailed discussion of the strengths and weaknesses of the various approaches are beyond the scope of the current paper, it is important to note the general distinction that is made between 'threshold' and 'non-threshold' toxic effects.

4.1 Toxic Effect With a Threshold

For some toxic effects, it can be assumed on the basis of mechanistic knowledge and dose–response information that a level of exposure exists below which no adverse health effects will occur – this might, for example, apply to proximal renal tubular cell loss in response to exposure to a xenobiotic that is excreted via the urine. In such instances, it is possible to use the dose that has been shown experimentally to represent either a no-observed-effect-level (NOEL) or no-observed-adverse-effect (NOAEL) level, to derive a daily or weekly intake or exposure that would be expected to be without effect in humans (see Section 8).

In the UK, such advisory standards may, depending on circumstance, be termed 'acceptable daily intakes' (ADIs) or 'tolerable daily intakes' (TDIs). In cases where experimental studies have failed to identify a NOEL/NOAEL but there is a strong mechanistic basis for believing that a threshold does exist, the level at which the lowest effect or adverse effect (*i.e.* the LOEL or LOAEL) was seen experimentally may be used; in such instances an additional uncertainty factor is applied to reflect the greater degree of uncertainty as to where the threshold level is situated. In some countries, a slightly different approach is applied, the so-called bench mark dose (BMD) method. In this, the actual experimental dose–response curves are mathematically modelled to determine a lower confidence bound for a dose equating to some specified response level – the BMD – usually selected at a response rate of between 1 and 10%, with a 95% lower confidence bound to ensure the value derived is conservative. The BMD, rather than the relevant NOEL/NOAEL/LOEL/LOAEL, is then used with appropriate uncertainty factors to define a reference dose (RfD). Known or predicted exposures can then be compared with the derived advisory standard to decide if the margin is adequate for any given exposure scenario.[15,17]

4.2 Toxic Effect Without a Threshold

Two types of toxic response are commonly believed not to have a threshold below which the possibility of an adverse effect can be discounted: mutagenicity and genotoxic carcinogenicity. Clearly, the deliberate exposure of humans to chemicals with such properties would be unacceptable and, wherever possible, such chemical usage should be banned (with the obvious exception of a pharmaceutical where the balance between risk and potential clinical benefit might be such as to warrant its use). However, particularly in environmental situations, humans may be inadvertently or unavoidably exposed to chemical mutagens and genotoxic carcinogens (of either 'natural' or anthropogenic origin). In such cases, particularly in the USA, mathematical models are used to quantify the expected level of response using the results of high-dose animal experiments to predict low-dose human exposure situations. Examples include Multi-hit, Weibull and multi-stage models, with the US Environmental Protection Agency (EPA) adopting a linearised multistage model (LMS) as a default approach in the absence of overriding pharmacokinetic, metabolic or mechanistic data. Such models allow derivation of a 'virtually safe or tolerable dose' (VSD, sometimes referred to a reference specific dose, RSD) based on an arbitrary level of risk (*e.g.* 1 in a million extra cancer cases on a lifetime exposure basis).

The robustness and predictivity of such modelling is open to considerable debate, and the use of such quantitative approaches to non-threshold risk assessments has yet to find much favour in the UK and Europe. Rather, approaches such as deriving an index dose (representing a maximum risk of, for example, 1 in 10^{-4} for a particular source of exposure) coupled with use of ALARA (as low as reasonably achievable) or BATNEEC (best available technology not entailing excessive costs) have found favour. Such an approach is used, for example, by the Environment Agency in the derivation of Soil Guideline Values for contaminated land for non-threshold chemicals.[13,15,17,18]

Looking ahead, it can be expected that the theoretical basis and practice of risk assessment will be heavily influenced by the ongoing developments in a number of quarters. For example, through greater integration of novel biomarkers to provide improved measures of effect and exposure, and information on the potential susceptibility of exposed populations; wider use of physiologically based pharmacokinetic (PBPK) models, that could be used to provide information to replace the current arbitrary default assumptions and to improve dose–response modelling and interspecies extrapolations; and use of probabilistic statistical techniques, such as Monte-Carlo analysis, that allow consideration of a range of distributions for the various input parameters to the risk assessment model, to produce a clearer and more realistic representation of the risk being assessed than current models which tend to use single (extreme) values.[15,19] Finally, for the first time, we are approaching the point at which we can realistically hope to see the development of models with the power to predict the toxic potential of mixtures of chemicals. Such models are, for example, currently in the early stages of development for endocrine disrupting activity.[20]

5 Role of Epidemiology

On first consideration, it would seem that the best way to measure the impact of chemicals, or any other 'stressors', on human health would be to assess the impacts in actual exposed human populations. Epidemiology provides the methods to do just that. There are, however, considerable problems in the conduct, analysis and interpretation of such studies[14] and while this approach is – in theory at least – applicable to existing exposure scenarios, it cannot of course be applied prospectively to situations where a novel or as yet unreleased substance is being assessed for potential human health impacts. In instances where a substance has already been used in industry, occupational studies can be informative regarding the assessment of risk to the general population from generalised environmental exposure; that is, if an occupational cohort has been investigated (normally at exposures much higher than would occur in the environment) then findings from such studies can be very informative in the overall assessment of risks associated with environmental exposure scenarios.

Epidemiological investigations of environmental influences on human health are mostly carried out using observational study designs; the three most common in which observations are recorded in individuals are cross-sectional, cohort, and case-control studies.[21] Cross-sectional studies describe the frequency of the disease of interest in a population at a particular period of time, and the variation is defined in terms of personal characteristics, time, place, etc., and – where available – relevant exposure metrics. Such studies provide a 'snap-shot' of a population and are useful for generating hypotheses. Cohort studies follow a group of individuals who have particular characteristics in common and observe the development of disease over time. The rate at which the disease develops in the exposed group in the cohort is compared with the rate in the non-exposed group, or in a standard group in the local or general population. In case-control studies, individuals with a given disease (the *cases*) are compared with a group of individuals without the disease (the *controls*). Information on past exposures to the chemical(s) or other risk factor of interest is then obtained for both cases and controls and compared – giving an '*odds ratio*' of disease associated with a particular exposure.

Where individual level studies are infeasible or impractical, group-level studies may be undertaken, but such 'ecological' studies are generally regarded to be of weaker design than individual level studies because necessary inferences about characteristics and exposures, for example, have to be made at the group level that may not pertain at the individual level – the so-called *ecological fallacy*.[21]

Two major potential problems in conducting epidemiological studies are selection bias and confounding.[22] Selection bias occurs when intervention/exposed groups and comparison/control groups differ in characteristics that affect the study outcome. This can occur just by chance or because an important subgroup is left out, either deliberately or by accident. If the differing characteristics have independent effects on the observed outcomes, they can confuse the interpretation of the study by creating outcomes that are not due to the intervention/exposure being assessed – this is called confounding. For a characteristic to be a confounder in a particular study it must be both causally related to the outcome and have a distribution that is different in the groups being compared. Steps are normally taken in epidemiological studies to prevent or adjust for important confounders, but sometimes this cannot be done, or done fully, which results in the potential for 'residual confounding' to create uncertainty in the interpretation of the results. Moreover, under or overcorrection for confounders can cause problems by either falsely weakening or strengthening a true association.

In studying the effects of low-level environmental pollutants, risks are likely to be small so that the possible unmeasured or residual confounding in the observational data is always a possibility to be considered. For example, socio-economic status and social deprivation are known to be linked to many health issues through complex interactions.[23] If, say, in a study on health effects of incineration the study population (located close to the incinerator) is more deprived than the control population, which is often the case, then correction for confounding is extremely important and even if a result persists after correction there may be doubt about the possibility of residual confounding especially if the outcome is or could be associated with the confounder. A good example is the large study by Elliott et al.[24] on cancer incidence near municipal solid waste incinerators in Great Britain. This found evidence for residual confounding near waste incinerators which, in the authors' words, seemed to be a likely explanation for the increased incidences of stomach, lung and all cancers combined, and also to explain at least part of the excess of liver cancer. Liver cancer is strongly related to deprivation and it was not possible to determine that the observed effect was not an artefact of incomplete or inappropriate correction for confounding.

6 Use of Biomarkers for Exposure and Risk Assessment

Biological and biochemical markers have long been used within clinical medicine in the diagnosis of diseases. While these are known to be useful in assessing risk to individuals and their likely response to clinical interventions, such markers have not found wide application in population screening.[25] However, biomarkers do have great potential in epidemiological and toxicological research to inform any assessment of risk posed by, and the possible consequences of, exposure of organisms to chemicals. Simplistically, three main types of biomarkers can be differentiated[26]: biomarkers of exposure, biomarkers of response or effect, and biomarkers of susceptibility.

6.1 Biomarkers of Exposure

These may further be defined as relating to either *internal* or *effective* dose. Internal dose biomarkers give an indication of the occurrence and extent of exposure of an individual organism and, thus, the likely concentration of parent compound or metabolite at a target site. Examples are urinary methylhippuric acid as a marker of short-term exposure to xylene in an occupational setting, and X-ray analysis of bone as a measure of long-term lead exposure. Biomarkers of effective dose relate to the true extent of exposure of what is believed to be the target molecule, structure or cell; for example, measurement of DNA or protein adducts in the blood as markers of exposure to alkylating agents. Biomarkers of exposure, where available, are preferable to simply measuring levels of the substance in the environment as they take account of the important influence of absorption, metabolism and distribution at the organism level.

6.2 Biomarkers of Response or Effect

These biomarkers often relate to measurement of pathological or biochemical changes indicative of biological (toxic) responses, but clinical or behavioural observations are also included in this category. There is increasing interest in the use of non-invasive techniques (*e.g.* measurements in urine or exhaled air) rather than blood or tissue samples. Further development of biomarkers of toxicity is required, for example in relation to early (preferably reversible) markers of lung damage, for which one such (a low-molecular-weight protein secreted by Clara cells) has already been identified.[27]

6.3 Biomarkers of Susceptibility

Biomarkers of susceptibility indicate whether an organism may behave differently from another organism in response to exposure to an external stressor. Particularly important sources of variability are polymorphism in metabolic capabilities (*e.g.* phenotypes for N-acetylation or glutathione S-transferase enzymes) and the presence of particularly susceptible genes for particular conditions (*e.g.* RB1 for retinoblastoma).

While this simple categorisation of biomarkers can be helpful in understanding their use and potential application, it is important to realise that there often exists a continuum between exposure and effect and a spectrum of biomarkers within this continuum. For example, the carboxyhaemoglobin formed after exposure to carbon monoxide is itself an *effect* of exposure but subsequent measurement of blood carboxyhaemoglobin (or exhaled carbon monoxide – which is in equilibrium with carboxyhaemoglobin) is also a direct measure of *exposure*.[26]

Biomarkers may be particularly useful in determining dose–response relationships, particularly if the biomarker can be measured at levels below which frank toxic responses can be confidently identified, and also potentially offer a means of facilitating cross-species extrapolations.[28] When used in risk assessment, biomarker data may allow for the replacement or refinement of default assumptions used in risk assessments. For example, it has been estimated that the risk of cancer from dichloromethane may vary by, on average, 25% if differences in frequency of glutathione 6-transferase Type 1 genotype (GSTT1) are included using a combined Monte-Carlo simulation and physiologically based pharmacokinetic (PBPK) model.[28]

Use of validated biomarkers can facilitate biologically based risk assessments, but to date there are few examples of their application in quantitative assessments.[28] Biomarkers of exposure and effect, in particular, can play an important role by informing the risk assessment process, and may prove particularly valuable in facilitating monitoring of the effectiveness of any risk management solutions; for example, by demonstrating if adequate levels of occupational protection are being achieved. In addition to the role of biomarkers in investigating the toxicity of a chemical, recent technological developments in the field of molecular biology and technology (*e.g.* genomics; see Section 7) offer very real possibilities of developing a range of biomarkers that are likely to be more sensitive, predictive and easier to apply than many of the types traditionally used. The wider use of biomarkers, however, requires risk assessors to become more adept at applying reliable measures of exposure or effect (including validated biomarkers) in regulatory assessments, and a greater willingness to move away from reliance on current default mathematical models.[28]

7 Applications of Molecular Biology

Increasingly, a wide range of molecular biological techniques, such as genomics, proteomics and metabolomics, are being applied to improve understanding of the toxicity profile of chemicals and, thereby, inform on the potential risk they may pose. Perhaps the currently most developed of these new approaches is the marriage of functional genomics with bioinformatics to characterise the effects and mechanisms of action of known or suspected toxicants, generally referred to as 'toxicogenomics'.

The most frequently applied tools are DNA microarrays and DNA chips, which allow the simultaneous monitoring of expression of literally hundreds or thousands of genes in cells exposed to chemicals *in vitro* or in tissues taken from exposed animals. Genes studied may include those coding for various Phase I and II metabolising enzymes and functionally related genes that may be pharmacological or toxicologically relevant. The information obtained can then be compared with previously established patterns of response or with the biochemical and pathological findings from traditional toxicity studies, so as to identify expression patterns that may be representative of adverse outcomes.[29] As such knowledge grows, this may allow the interpretation of toxicogenomic data prior to conducting more time-consuming traditional toxicological methods and, through cross-species comparison of responses, improve the predictive accuracy of extrapolations from *in vitro* to *in vivo* and *ex vivo* animal models to humans.[30]

A concrete example of genomic technology improving understanding of risks posed by chemicals is in relation to the study of endocrine disrupting chemicals. Many natural substances (*e.g.* the mammalian hormone oestradiol, and plant constituents such as the soy isoflavone genistein) or of anthropogenic origin (*e.g.* Bisphenol A) are hormonally active, but there is considerable uncertainty as to the relative risk posed to humans as a result of environmental exposures. Recent findings using molecular biological techniques have demonstrated that, at least for oestrogen receptor (ER)-responsive genes, the pattern of gene expression appears to be very similar, irrespective of the type of chemical. For example, in MCR-7 cells expressing the ERα subtype, the gene stimulation profile is similar, though not identical, for a natural steroid, oestradiol, and the synthetic chemical Bisphenol-A, with both affecting a series of ER-dependent genes as well as some unique targets.[31] The *in vivo* expression of a panel of 179 genes in the uterus of immature Alpk:ApfCD-1 mice has also been shown to be similar for equipotent dosages of oestradiol, a synthetic steroid diethylstilbestrol (DES) and genistein, notwithstanding some differences in magnitude of effect.[32] Such findings suggest that oestrogenically active substances of different origin may all have a common potential to elicit effects and, thus, it may be appropriate to undertake risk assessments for natural as well as synthetic chemicals possessing such properties.

Other workers are attempting to assess the level of risk posed by multiple chemical exposures; for example through measurement of DNA adducts in human umbilical cord tissue to assess the effects of exposure to multiple chemicals.[33] Also, work is underway to investigate the application of toxicogenomics to predictive toxicology. For example, Steiner *et al.*[34] have used traditional toxicity data, combined with knowledge of hepatic gene expression profiles for a series of chemicals with known mechanisms of toxicity, to 'train' mathematical models (in this case support vector machines) to generate classification rules allowing the identification of a probe subset of genes to permit classification of compounds as hepatoxic and non-hepatotoxic. For a test set of 63 compounds, almost 90% of the toxic group were correctly classified as toxic and no false-positives were generated, although some false negatives were produced.

Looking to the longer term, Simmons and Portier[30] have proposed that completion of the sequencing of the human genome and increasing knowledge of genetic polymorphisms, coupled with improved mathematical models representative of underlying physiological, biological and biochemical processes, may allow toxicogenomics to be used to directly define what would constitute a reasonable exposure limit for chemicals present in the environment. However, before such a situation becomes a realistic and practical possibility, a number of significant challenges must be overcome.

There is an urgent need to extend the scope of the toxicogenomic datasets to include a wider range of chemicals with different modes of action over a range of dosages and for a range of species and strains, and to ensure that the information gathered is thoroughly validated; bioinformatics techniques and predictive models must be further developed and validated; approaches must be developed that more readily allow separation of gene changes associated with toxicity from those that are of only minor importance or are actually adaptive or protective in nature – that is to say, current techniques cannot readily distinguish between causative, adaptive or incidental changes; limitations in our knowledge of the genome means that it is often not possible to interpret the meaning of changes in gene expression; there remains the perennial problem for *in vitro* test systems of limitations in the ability to investigate interactions between chemicals or with other environmental factors, and the influence of physiological processes and metabolism on the ultimate expression of toxicity for a given compound; ability to extrapolate to the *in vivo* situation is also limited because a chemical may act through multiple mechanisms, and effects may be dependent on dose, time and duration of exposure and the organism's phenotype and life stage; and there is a need to consider the significance of extra-genomic changes, such as the expression of transcription factors and the possible role of changes in upstream signalling and processing (which will require integration of the likes of proteomic and metabolomic techniques).

A symposium held by the UK Committee on Toxicity of Chemicals in Food, Consumer Products and the Environment[35] concluded that further research and validation was necessary before toxicogenomic and proteomic techniques could be considered for routine use in regulatory toxicological risk assessment. Similar observations were made at a recent IPCS workshop.[36] Until such issues are resolved, however, risk assessors and regulators can expect to be faced with major challenges; for example, on the interpretation of reports of altered gene profile in the absence of causes of concern from traditional toxicity studies, or identification of potentially susceptible human sub-populations. This will necessitate making decisions as to the implications, if any, of such data when establishing acceptable occupational or environmental exposure limits, or even the possibility of legal liability claims based upon gene expression data. It can be expected that, for the foreseeable future, the use of toxicogenomic data as supportive information will need to be approached on a case-by-case basis until its predictive value is firmly established.

8 Uncertainties in Human Health Risk Assessment and the Role of Expert Committees

There are many uncertainties in the process of establishing whether a risk to human health is associated with exposure to a particular chemical – and, if so, what a 'safe' level of exposure might be. Epidemiological studies are subject to the problems outlined above, and whenever an association is detected, very often the question remains whether that association is causal or spurious. Ecological epidemiological studies alone are not able to prove causality – nonetheless, where the toxicology is sufficiently well established, occupational epidemiology studies have been used to help in determining no-effect levels for humans in an environmental context; as done, for example, by the Expert Panel on Air Quality Standards in producing standards for benzene and 1,3-butadiene in ambient air.[37,38]

The use of experimental animal data to determine toxicity is subject to a number of uncertainties, particularly regarding read-across from animals to humans (inter-species variation) and the shape of the dose–response curve, especially at low dose levels. Where experimental studies are able to ascertain a NOEL, this is used in combination with an uncertainty factor to establish an equivalent no effect level for humans. If only a LOEL can be determined, an additional uncertainty factor is multiplied in. Factors to adjust for exposure period and to allow for intra-species differences (individual susceptibilities) are also applied to derive a 'safe' level for human exposure. Default approaches and numerical values for such uncertainty factors have been agreed by a number of different bodies.[3,39] The derivation of a TDI for phenol and an ADI for the herbicide linuron are illustrative examples of the process (see Box 1).

Other – often significant – uncertainties in the risk assessment process relate to the estimation of exposure. Comprehensive high quality exposure data are rare in occupational settings and even rarer in the environmental context because monitoring is often sparse. Modelling is therefore often used to estimate exposure, but can be unreliable and difficult to validate.[3,12] There is frequently uncertainty with regard to missing or incomplete information to define exposure and dose (scenario uncertainty), the necessary parameters (parameter uncertainty), and with regard to gaps in scientific theory on causal inference (model uncertainty).

Because of the range of uncertainties involved in health risk assessment for environmental chemicals, it is common practice to subject such evaluations either to extensive peer review or to use an expert committee to make the final judgements on particular outcomes and conclusions. Thus the European process for chemical risk assessment, as defined for example in the Technical Guidance Document (TGD) on Risk Assessment of Chemical Substances following European Regulations and Directives[40] involves cross-agency and cross-national inputs into the assessment and interpretation of data. In the UK (and elsewhere) there are a number of expert advisory panels, generally composed of a Civil Service secretariat, representatives of relevant departments, independent (academic) scientists and, in some cases, other stakeholder organisations. These are set up to undertake and/or evaluate various aspects of the chemical health risk assessment process – depending on the nature and application of the assessment – e.g. the Committee on the Medical Aspects of Air Pollutants (COMEAP), the Advisory Committee on Hazardous Substances (ACHS), the Committees on Toxicity, Mutagenicity and Carcinogenicity of Chemicals in Food, Consumer Products and the Environment (COT, COM, COC) and the Expert Panel on Air Quality Standards (EPAQS). These committees generally work on a 'consensus approach' and are mostly employed to identify best practice, set criteria and procedures and, as appropriate, determine safe levels (standards or guidance) for

Box 1

> **Using experimental toxicity data to derive human intake standards**
>
> *Phenol*
>
> The UK Committee on Toxicity of Chemicals in Food, Consumer Products and the Environment (COT) assessed the toxicological data on phenol with the objective of establishing a soil guideline value. A critical study in rats was identified that showed a NOEL of 70 mg/kg (bw)/day. It was considered that standard uncertainty factors of 10 for extrapolation from rodent data and 10 for variability within the human population were appropriate to apply to this NOEL, leading to the derivation of an oral TDI of 0.7 mg/kg (bw)/day.
>
> *Linuron*
>
> When the UK Advisory Committee on Pesticides (ACP) considered the data on the substituted urea herbicide, linuron, evidence of potentially important effects were noted in long-term studies in rats. Even at the lowest dose tested (25 ppm in diet, equivalent to 1.3 mg linuron/kg (bw)/day), a reduction in incidence of pituitary tumours was apparent. While a reduction in tumour incidence might, at first sight, be considered an unlikely cause for concern, this change was attributed to altered hormonal status in the animals and hence was considered to be of potential significance to human health. A factor of 10 was applied for interspecies extrapolation and another factor of 10 for variability within the human population – plus an additional factor of five because of the use of an experimental LOEL rather than a NOEL, thus giving an overall uncertainty factor of 500. On this basis, an ADI of 0.003 mg/kg bw/day was calculated[39].

levels of contaminants in environmental and other media. Good examples of standard setting for environmental pollutants in the UK are the work of EPAQS (*vide infra*) and the derivation of guidelines for soil contaminants undertaken by the Environment Agency.[13]

9 The Changing Face of Chemical Regulation in Europe

Against a background of increasing public and political disquiet about the adequacy of protection from environmental pollution offered to humans and wildlife, in February 2001 the European Commission (EC) published a White Paper[41] that signalled a major upheaval of chemical regulation within the European Union.

Since the implementation in 1979[42] of the sixth amendment to the 1967 directive on dangerous substances,[43] there had been a major division in approach to chemical regulation within Europe: approximately 100,000 so-called 'existing chemicals', *i.e.* those on the market prior to 1981, required only very limited hazard (toxicity) and risk assessment; but increasingly demanding testing as their annual production volumes increase stepwise beyond 10 kg is required for the 2700 'new chemicals' that have been marketed since 1981.[41]

The 2001 White paper[41] proposed a unified system, REACH (Registration, Evaluation and Authorisation of CHemicals), for all chemicals produced at annual volumes of greater than 1 tonne not covered by specific, more demanding legislation – as in the case of pharmaceuticals and agrochemicals. REACH would also significantly simplify legislation by replacing 40 existing Directives. The Commission's proposal has now been refined as a result of consultation with individual Member States and public and stakeholder organisations. At the time of writing, a revised REACH proposal[44] is in its review stage at the European Parliament. This differs from the original 2001 proposal in a number of ways, in particular in the creation of two types of 'Evaluation' and addition of a 'Restriction' option (see Box 2). While the ultimate shape of legislation is not yet certain, the expectation is that the final form will be similar to the 2003 proposal.

A key element of the REACH proposal is the stepwise increase in toxicity data requirements as a chemical's market or production volume increases. At the low volume of 1–10 tonnes/annum (where test requirements are limited to *in vitro* assessments of mutagenicity, corrosivity and irritation, a daphnia acute toxicity test and an *in vivo* rodent sensitisation study) particularly, this represents a reduction in testing load compared to that currently required for a new chemical. However, for high volume-chemicals, the amount of data required on both mammalian and eco-toxicity may be extensive. The EC proposes to limit the extent of vertebrate testing needed by the development and application of alternatives, such as quantitative structure–activity relationship (QSAR) and *in vitro* models. However, the extent to which this will be practicable within the envisaged time scales for implementation of REACH is uncertain.[45,46]

REACH will have a major impact on chemical risk assessment practices in Europe, in particular with respect to abolition of the difference in approach to 'existing' and 'new' chemicals and the change from the current system where government authorities are responsible for risk assessment, to the producer or importer of a chemical becoming responsible for ensuring the adequacy of risk assessment undertaken. The producer/importer will also be obliged to make toxicity and exposure data and safe-handling advice available to the authorities, customers along the entire supply chain, and the public. In addition, specific permission (authorisation) will be needed for the continued use of the most dangerous chemicals (see Box 2) for which it will be necessary to demonstrate the absence of suitable alternatives and/or a compelling socioeconomic need.

It is currently thought that the REACH regulations will come into force during 2006 or, at the latest, 2007, but implementation will be staggered over at least 11 years, with the first registration deadline of 3 years applying only to chemicals produced at very high volumes (>1000 tonnes/annum) and those produced at greater than 1 tonne/annum that possess highly toxic or persistent properties. Varying assessments of the health, economic and social impacts of the proposed changes have been made. For example, total costs for hazard testing have been estimated at €2.3–8.7b[45] with a further downstream cost to industry of €2.8–5.2b. WWF[47] estimated that REACH could cost industry up to €23.6b but would result in health cost

Box 2

> ### Elements of REACH
>
> ***Registration*-EACH:** Substances produced or imported in quantities of one tonne or more per year (with limited exceptions) are to be registered in a central database, administered by a new European Chemicals Agency (ECA). Information is required on physical and hazardous properties, uses of the substance, risk management measures and any proposed further testing. Extent of information required is driven by production and market volume.
>
> **R-*Evaluation*-ACH:** Two forms are proposed: Dossier Evaluation – where Regulatory Authorities may check compliance and accuracy of the dossiers submitted, or review a dossier to decide if proposals for further toxicity testing are warranted; and Substance Evaluation – where information may be evaluated in cases of suspicions that there may be a risk to human health or the environment. The conclusions of these evaluations may lead to initiation of either the authorisation or restriction mechanisms.
>
> **RE-*Authorisation*-CH:** Authorisation is required for substances of very high concern on the basis of their biological properties. Examples include chemicals that are persistent, bio-accumulative and toxic (PBT), very persistent and very bio-accumulative (vPvBs) or that are carcinogenic, mutagenic or toxic to reproduction (CMR) or that posses toxic properties of equivalent concern (as yet ill-defined but including those with endocrine disrupting activity). Authorisation is granted for particular stated uses only where the risks can be adequately controlled or where the socio-economic grounds outweigh the risk and no technological alternatives or substitutes exist.
>
> **REA-(*Restriction*)-CH:** Restrictions on the manufacture, use and/or placement on the market of a substance (up to and including a complete ban) may be imposed on any substance that, while not meeting the definition of very high concern (as required for the authorisation process), is considered to pose an unacceptable risk.

savings of over €74.9 billion over 17 years by reducing exposure to harmful chemicals through their withdrawal or substitution. A report for the EC[48] identified potential savings for occupational health alone of €27–54b. In 2003, the EC estimated the total health benefit to be of the order of €50b over 30 years,[49] and a more recent assessment[50] has suggested that costs to industry may be significantly less than originally anticipated.

The extent to which the hoped-for positive balance between cost and benefit is realised is as yet uncertain. However, it is clearly the case that the introduction of REACH will place a considerable onus on those scientists and regulators with expertise in chemical hazard and risk assessment to ensure that the system is adequately supported and that alternative test models are developed to limit the use of live animals, while quantification of the actual health benefits achieved will present significant challenges to epidemiologists and public health specialists over the coming decades.

10 Future Perspectives

The risk to human health posed by exposure to chemicals in the environment is likely to remain an important concern for both the general public and for the scientists and regulators who have to assess and manage that risk. Historically, the process of attempting risk assessments for the tens of thousands of substances on the market has been painfully slow, a major problem in itself, and there are increasing concerns over the animal welfare issues associated with the toxicological testing of these substances. Further problems and uncertainties attach to the issues of exposure to mixtures and current assumptions about dose–response relationships. As explained in this chapter, epidemiology is limited in its application here and complicated in its interpretation.

Although the application of environmental chemistry and toxicological principles have allowed progress to be made in the assessment of health risks from environmental chemicals, it is clear that current procedures have limitations when faced with the sheer scale of the issues and the current degree of public concern. Introduction of REACH into Europe will have a dramatic impact on the way chemicals are regulated, although only time will tell whether this major overhaul of the regulatory process will actually deliver the desired benefits. Combined with present and predicted developments in alternative test methods, including *in vitro* and *in silico* (computer-based modelling; e.g. quantitative structure–activity relationships) approaches, and the application of molecular biology, it is to be expected that chemical risk assessment in the future will differ significantly from current procedures. Also there is the arrival of 'green chemistry' and the increasing consideration of different approaches to the provision of 'services' that may lead, at least in some cases, to the substitution of chemical products by non-chemical processes, while the new 'nanotechnologies' are spawning a multitude of innovations but also a series of pertinent questions and issues that need to be addressed. Endocrine disruption, meanwhile, will continue to be a thorn in the side of risk assessors and regulators because of the difficulties in testing for and interpreting adverse affects, not to mention problems posed by the purported 'low-dose effects' of endocrine disrupters, that will inevitably further complicate and delay the development and application of standard approaches.

Although very often the news media are accused by scientists of misrepresenting and exaggerating the risks posed by chemicals in the environment, it is clear that the public's perception of risk issues is important, and over the past few decades the public's influence on the political agenda and on policy development has grown steadily. Risk assessment procedures are now more accessible, and expert advisory committees typically have to ensure that their papers are available for open scrutiny, for example by placing them on the internet. Many committees already include a 'lay member' to represent the public; it is conceivable that in the future there will be much more extensive and thorough public involvement in the chemicals risk assessment process and that 'expert judgement' itself will include societal inputs.

References

1. P.T.C. Harrison, in *Pollution Causes, Effects and Control*, R.M. Harrison, (ed.), Royal Society of Chemistry, Cambridge, UK, 2001, 500.

2. V. Herbert, *Agrichemical and Environmental News*, 2001, **177**, Available [May 2005] at: http://www.aenews.wsu.edu/.
3. WHO, *Principles for the assessment of risks to human health from exposure to chemicals, International programme on chemical safety, environmental health criteria*, 210, World Health Organization, Geneva, Switzerland, 1999.
4. Risk Assessment and Toxicology Steering Committee, *Developing new approaches to assessing risk to human health from chemicals*, cr1, MRC Institute for Environment and Health, Leicester, UK, 1999.
5. OECD, *OECD guidelines for the testing of chemicals* (up to and including Thirteenth Addendum), Vols, 1 and 2, Organisation for Economic Co-operation and Development, Paris, France, 2000.
6. IEH, *Environmental oestrogens: Consequences to human health and wildlife*, Assessment A1, MRC Institute for Environment and Health, Leicester, UK, 1995.
7. IEH, *Assessment on the ecological significance of endocrine disruption*, Assessment A4, MRC Institute for Environment and Health, Leicester, UK, 1999.
8. WHO, *Global Assessment of the State-of-the-science of endocrine disruptors*, International Programme on Chemical Safety, WHO/PCS/EDC/02.2, World Health Organization, Geneva, Switzerland, 2002, Available [May 2005] at; http://www.who.int/ipcs/publications/new_issues/endocrine_disruptors/en/.
9. CSTEE, *Opinion on human and wildlife health effects of endocrine disrupting chemicals, with emphasis on wildlife and on Ecotoxicology test methods*, Scientific committee on toxicity, ecotoxicity and the environment, European commission, directorate general for Consumer Policy and consumer health protection, Brussels, Belgium, 1999.
10. OECD, *Current Activities in the OECD Test Guideline Programme on Endocrine Disrupters*, Organisation for Economic Co-operation and Development, Paris, France, 2005, Available [May 2005] at; http://www.oecd.org/document/42/0,2340,en_2649_34377_2348650_1_1_1_1,00.html.
11. M. Balls, A. Bogni, S. Bremer, S. Casati, S. Coecke, C. Eskes, P. Prieto, E. Sabbioni, A. Worth, V. Zuang, M. Barratt, B. Blaauboer, P. Botham, R. Combes, J. Doehmer, J. Fentem, M. Liebsch, and H. Spielmann, *ATLA-Alternatives to laboratory animals*, 2002, **30** (Suppl 1), 1.
12. Risk Assessment and Toxicology Steering Committee, *Exposure assessment in the evaluation of risk to human health*, cr5, MRC Institute for environment and health, Leicester, UK, 1999.
13. Defra and Environment Agency, *Contaminants in soil: Collation of toxicological data and intake values for humans*, R&D Publication CLR 9 , environment agency R&D Dissemination Centre c/o WRc, Swindon, UK, 2002.
14. G.A. Ames, in *Toxicology and risk assessment, principles, methods and applications*, A.M. Fan and L.W. Chang (eds.), Marcel Dekker, New York, 1996, 559.
15. M.L. Gargas, B.L. Finley, D.J. Paustenbach, and T.F. Long, in *Chlorinated organic micropollutants*, R.E. Hester and R.M. Harrison (eds.), Royal Society of Chemistry, Cambridge, UK, 1996, 89.
16. W.C. Roberts and C.O. Abernathy, in *Toxicology and risk assessment, principles, Methods and Applications*, A.M. Fan and L.W. Chang (eds.), Marcel Dekker, New York, 1996, 245.

17. Risk Assessment and Toxicology Steering Committee, *Risk Assessment Approaches used by UK Government for evaluating human health effects of chemicals*, cr2, MRC Institute for Environment and Health, Leicester, UK, 1999.
18. D.P. Lovell and G. Thomas, in *Food Chemical Risk Analysis*, D.R. Tennant (ed.), Blackie Academic and Professional, London, UK, 1992, 57.
19. IEH, *Food risk assessment: Probabilistic approaches to food risk assessment*, FORA 1, MRC Institute for Environment and Health, Leicester, UK, 2000.
20. J. Payne, N. Rajapakse, M. Wilkins, and A. Kortenkamp, *Environ. Health Perspect.*, 2000, **108**, 983.
21. L. Rushton and P. Elliott, *Bri. Med. Bulle.* 2003, **68**, 113.
22. M. Mamdani, K. Sykora, P. Li, S.L. Normand, D.L. Streiner, P.C. Austin, P.A. Rochon, and G.M. Anderson, *BMJ*, 2005, **330**, 960.
23. D. Wanless, *Securing good health for the whole population*, Final Report, HMSO, London, UK, 2004, Available [Feb. 2005] at; http://www.hm-treasury.gov.uk/consultations_and_legislation/wanless/consult_wanless04_final.cfm.
24. P. Elliott, G. Shaddick, I. Kleinschmidt, D. Jolley, P. Walls, J. Beresford, and C. Grundy, *Br. J. Cancer*, 1996, **73**, 702.
25. IEH, *Diet-Gene interactions: Characterisation of risk*, FORA 3, MRC Institute for Environment and Health, Leicester, UK, 2001.
26. C.J. Waterfield and J.A. Timbrell, in *General and Applied Toxicology*, B. Ballantyne, T.C. Marrs, and T. Syversen (eds.), McMillan Oxford, UK, 2000, 1841.
27. IEH, *Approaches to predicting toxicity from occupational exposure to dusts: An appraisal of low-toxicity dusts*, Report R11, MRC Institute for Environment and Health, Leicester, UK, 1999.
28. WHO, *Biomarkers in risk assessment: Validity and validation*, International programme on chemical safety, environmental health criteria 222, World Health Organization, Geneva, Switzerland, 2001.
29. V.S. Chan and M.D. Theilade, *Clin. Toxicol.*, 2005, **43**, 121.
30. P.T. Simmons and C.J. Portier, *Carcinogenesis*, 2002, **23**, 903.
31. D.W. Singleton, Y. Feng, Y. Chen, S.J. Busch, A.V. Lee, A. Puga, and S.A. Khan, *Mol. Cell Endocrinol.*, 2004, **221**, 47.
32. J.G. Moggs, J. Ashby, H. Tinwell, F.L. Lim, D.J. Moore, I. Kimber, and G. Orphanides, *Environ. Health Perspect.*, 2004, **112**, 1137.
33. C. Mori, M. Komiyama, T. Adachi, K. Sakurai, D. Nishimura, K. Takashima, and E. Todaka, *Environ. Health Perspect. Toxicogenomics*, 2003, **111**, 7.
34. G. Steiner, L. Suter, F. Boess, R. Gasser, M.C. de Vera, S. Albertini and S. Ruepp, *Environ. Health Perspect.*, 2004, **112**, 1236.
35. COT, COM, and COC, *Joint statement on the use of toxicogenomics in toxicology*, COT/04/10; COM/04/S5;COC/04/S8 December 2004, Committees on Toxicity, Mutagenicity and Carcinogenicity of Chemicals in Food, Consumer Products and the Environment, London, UK, 2004, Available [May 2005] at; http://www.food.gov.uk/science/ouradvisors/toxicity/.
36. WHO, *Toxicogenomics and the risk assessment of chemicals for the protection of human health*, IPCS/Toxicogenomics/03/1, World Health Organization,

International Programme on Chemical Safety/International Labour Organization and the United Nations Environment Programme, 2003.
37. EPAQS, *Benzene*, Expert panel on Air Quality Standards, HMSO, London, UK, 1994.
38. EPAQS, *1,3-Butadiene*, Expert panel on Air Quality Standards, HMSO, London, UK, 1994.
39. IGHRC, *Uncertainty Factors: Their use in human health risk assessment by UK Government*, cr9, MRC Institute for Environment and Health, Leicester, UK, 2003.
40. ECB, *Technical Guidance Document (TGD) on Risk Assessment of Chemical Substances following European Regulations and Directives*, 2nd Edition, European Chemicals Eureau, Ispra, Italy, 2003, Available [June 2005] at; http://ecb.jrc.it/tgdoc.
41. EC, Strategy for a Future Chemicals Policy – REACH White Paper, COM(2001) 88 Final, European Commission, Brussels, Belgium, 2001.
42. EEC, Directive 79/831/EEC of September 1979 amending for the sixth time Directive 67/548/EEC on the approximation of the laws regulations and administrative provisions relating to the classification, packaging and labelling of dangerous substances, 1979, OJ L259, 15 October 1979.
43. EEC, Directive 67/548/EEC of 27 June 1967 on the approximation of the laws regulations and administrative provisions relating to the classification, packaging and labelling of dangerous substances, 1967, OJ L196, 16 August 1967.
44. EC, Proposal for a regulation of the European Parliament and the Council concerning the Registration, Evaluation, Authorisation and Restrictions of Chemicals (REACH), COM(2003) 644, European Commission, Brussels, Belgium, 2003.
45. POST, EU Chemical Policy, Postnote Number 229, Parliamentary Office of Science and Technology, London, UK, 2004.
46. IEH, *Assessment of the feasibilty of replacing current regulatory In Vivo Toxicity Tests within the framework specified in the EC white paper Strategy for and EU Chemicals Policy*, IEH Web Report W10, MRC Institute for Environment and Health, Leicester, UK, 2001, Available [May 2005] at; http://www.le.ac.uk/ieh/.
47. WWF, The Social Costs of Chemicals: – The cost and benefits of future chemicals policy in the european union, WWF-UK, Godalming, UK, 2003.
48. RPA, Assessment of the impact of the new chemicals policy on occupational health, risk and policy analysts Ltd, London, UK, 2003.
49. EC, Commission Staff Working Paper SEC(2003) 1171/3 – Regulation of the European Parliament and the Council concerning the Registration, Evaluation, Authorisation and Restrictions of Chemicals (REACH), establishing a European Chemicals Agency and amending Directive 1999/45/EEC and Regulation (EC) on Persistent Organic Pollutants – Extended Risk Assessment, COM(2003) 644 Final, European Commission, Brussels, Belgium, 2003.
50. KPMG, REACH – further work on impact assessment: A case study approach, executive summary, KPMG Business Advisory Services, London, UK, 2005.

Environmental Risk Assessment

LORRAINE MALTBY

1 What is Environmental Risk Assessment?

Risk assessment is the process of estimating the likelihood that a particular event will occur under a given set of circumstances. Hence, environmental risk assessment involves an analysis of information on the environmental fate and behaviour of chemicals in the environment (*i.e.* air, water and land) integrated with an analysis of information on their effects on human beings and ecological systems.[1] Ecological risk assessment is that component of environmental risk assessment which is concerned with the effects of chemicals on non-human populations, communities and ecosystems.[2] Risk assessment is an important decision-making tool that can be used to identify existing problems and to predict potential risks of planned actions. It is useful both for prioritising management and regulatory efforts and for evaluating the effectiveness of management actions that are implemented.

Central to any risk assessment process is the distinction between hazard and risk. Whereas hazard is the ability of a chemical to harm organisms, risk is the probability that harm will occur under a particular set of circumstances. A chemical may be extremely hazardous, but if there is no environmental exposure, it will not present an environmental risk. Risk assessment brings together information on exposure and effects, and both are of equal importance. The fate and behaviour of chemicals in the environment is the focus of Chapter 7 [3] and is therefore not discussed in detail here. Rather, the emphasis of this chapter will be on the effects of chemicals on ecosystems and their components and the use of this information to assess risk. The risks that chemicals in the environment pose to human health are discussed in Chapter 4.[4]

1.1 Prospective and Retrospective Risk Assessment

Prospective (or predictive) risk assessment predicts the likely consequences of releasing a chemical substance into the environment. It is used to help the risk manager and regulator decide whether a chemical substance should be registered for use, and if so, under what circumstances and with what control measures. Prospective risk assessments are usually generic rather than site-specific and, consequently, require the identification of the potential effects of a chemical on a large number of organisms in a variety of ecosystems. As it is not possible to study all potentially affected species and habitats, generic prospective risk assessments are based on indicator organisms and defined environmental compartments. For example, the standard exposure scenario used in the aquatic risk assessment of plant protection products in Europe is a static ditch containing a 30-cm depth of water overlying a 5-cm depth of sediment and the standard test organisms are an alga, an invertebrate and a fish.[5]

Retrospective risk assessments assess the effects of chemicals once they have been released into the environment and may be regional, local or site-specific. The focus of retrospective risk assessments is usually the effect of a single chemical or chemical source on selected ecosystem components. However, ecosystems are comprised of a large number of interacting species and the structure and functioning of ecosystems are the consequence of a variety of physical, chemical and biological factors acting sequentially or concurrently. Teasing out the effects of a single chemical from the multitude of natural and anthropogenic stressors operating on ecosystems is one of the major challenges of retrospective risk assessment.

1.2 The Risk Assessment Process

Assessing the risks that chemicals pose to the environment is a three-phase process (Figure 1). The process begins with a problem formulation phase in which the hazard is identified and the study is planned. This is followed by an analysis phase in which the exposure to, and effects of, the chemical are assessed. Finally, exposure and effects information are brought together to characterise risk. Risk assessments may be performed for different environmental compartments (*i.e.* aquatic, terrestrial, atmospheric) and a tiered approach is usually adopted. Lower tier risk assessments are based on limited data and reasonable worst-case assumptions, whereas higher-tier assessments are based on more realistic, but complex data sets. Moving through the tiers refines the assessment of exposure and effects and hence reduces uncertainty in the risk characterisation.

The purpose of the problem formulation stage is to determine whether a particular danger exists, and if so, whether the associated effects warrant further study or management action. If further studies are necessary, then the types of data required to characterize the risk are identified and appropriate assessment and measurement endpoints selected. Information used for hazard identification includes short-term or screening toxicity tests and reviews of existing information that characterise the potentially affected ecosystems and contaminants in question. In retrospective risk assessments, this phase should include the development of a conceptual model that identifies contaminant sources, biological receptors and the processes that link them. It should also include the identification of assessment endpoints (what is to be protected?), the selection of measurement endpoints (what is to be measured to determine exposure and effects?) and the determination of level of effect to be detected (how large a change is of concern?).

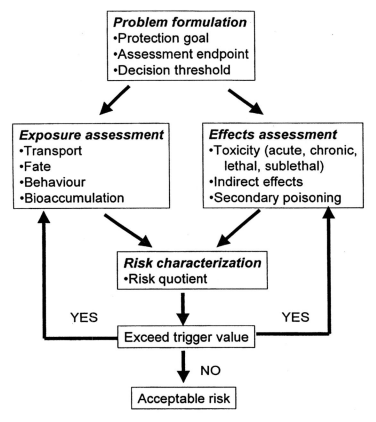

Figure 1 *Environmental risk assessment process*

The analysis phase includes measurement or prediction of the emission, transport, fate and behaviour of the chemical in the environment. The purpose of exposure assessment is to determine exposure concentrations and identify key environmental components for assessing risks. In prospective risk assessments, predicted environmental concentrations (PECs) are based on predicted emissions and use data and are derived using mathematical modelling. Mathematical modelling may also play a key role in retrospective risk assessment, but it is usually supported by chemical analysis of environmental media or ecological receptors. In addition to exposure assessment, the analysis phase characterises the relationship between exposure and effects in order to derive concentrations at which no adverse ecological effects occur (*e.g.* predicted no effect concentration, PNEC).

Effect assessments can be based on information obtained from single-species toxicity tests, microcosm or mesocosm studies, field studies, field surveys or population and ecosystem modelling. Lower-tier effects assessments are based on short-term effects data derived from acute toxicity tests with standard species (*i.e.* algae, *Daphnia*, fish, chironomid larvae). Acute toxicity is usually expressed as median lethal (effective) concentration (L(E)C50), which is the concentration required to kill (effect) 50% of the individuals exposed to the chemical in a given time period. Higher-tier effects assessments use data from acute and chronic toxicity tests conducted on a range of species. Chronic toxicity tests usually measure the sublethal effects of chemicals (*e.g.* effects on growth, reproduction, development) and toxicity is expressed as a no observed effect concentrations (NOEC). The NOEC is the test concentration immediately below the test concentration causing a statistically significant effect on the test endpoint (*i.e.* lowest observable effect concentration, LOEC). Higher-tier assessments move beyond the study of the effect of a chemical on individual organisms to consider population-level and community-level effects, including indirect toxic effects due to changes in trophic interactions and secondary poisoning.[6,7]

At the end of the analysis phase, information from the exposure and effects assessments are brought together to describe the nature and magnitude of risks posed by the chemical. Risk may be characterised as a quotient (*i.e.* toxicity exposure ratio (TER), PEC/PNEC ratio) or as a probability, and should include a consideration of the uncertainties inherent in the risk assessment process. As the purpose of risk characterisation is to provide information to risk managers and decision makers in a manner that is understandable and relevant to the decisions being made, the level of risk that is acceptable or which triggers further study, should be clearly defined in the problem formulation phase.

1.3 Uncertainty and Variability

Environmental risk assessments are subject to both variability and uncertainty. Variability is an inherent property of the system being investigated and includes interspecific, inter-population and inter-individual differences in exposure and response as well as temporal and spatial differences in the biotic and abiotic components of ecosystems. Variability cannot be reduced by increased information or measurement, but it can be characterised and its influence minimised by careful selection of study systems. Uncertainty is due to imprecise or incomplete knowledge and is a property of the relationship between the study system and the risk assessor. Uncertainty in risk assessment may be categorised as variable uncertainty or model uncertainty. Variable uncertainty arises from imprecise, inaccurate or inappropriate measurements and can be reduced by improving measurement techniques and study design. Model variability arises from an incomplete mechanistic understanding of the system under investigation and is a feature of both extrapolations and mathematical or conceptual models.

Uncertainty and variability inherent in the risk assessment process may mean that the true effects are larger than the estimated effects. Consequently, uncertainty and variability must be incorporated into the analysis of risk. Risk assessments are often based on information obtained on a small number of species, exposed to a single chemical under a given set of conditions for a limited period of time. This information is then used to estimate the generic risk of both short-term and long-term exposure of complex ecosystems exposed to multiple stressors; a process that incorporates many uncertainties. Risk assessment procedures take one of two approaches to addressing uncertainty. The first, commonly applied to deterministic risk assessments, is to apply uncertainty (or assessment) factors to either model parameters or outputs; the second is to analyse uncertainty using probabilistic approaches.

1.4 Deterministic and Probabilistic Risk Assessments

Deterministic risk assessments use fixed values to estimate toxicity (*e.g.* L(E)C50, NOEC) and exposure (*e.g.* PEC) and generate a single measure of risk, such as a risk quotient. In contrast, probabilistic risk assessments express results of exposure and/or effect assessments as probability distributions and it is these distributions that are used to generate a probability distribution of risk estimates.[8] Deterministic risk assessments are simple to perform and require relatively little data. They generally assume worst-case conditions and characterise risk using the lowest available toxicity estimate and the highest available exposure estimate. Consequently, deterministic risk assessments tend to be very conservative and may over-estimate risk. In contrast, probabilistic risk assessments require quantitative information on exposure and effects and therefore can only be applied to data-rich chemicals. However, probabilistic approaches have two major advantages over deterministic approaches: first, they provide quantitative information on the probability of an adverse effect occurring; second, they can include estimates of uncertainty and variability.

Major sources of uncertainty in risk assessment include those associated with extrapolating across: response endpoints (*e.g.* lethal, sublethal), individuals and species, study systems (*e.g.* laboratory test, mesocosm study, field site), spatial scales, temporal scales and geographical locations. For most chemicals there is limited data with which to predict ecosystem effects and in most cases only acute toxicity data are available. Extrapolating from single-species short-term toxicity data to ecosystem effects involves extrapolating from few species to many species, from short-term exposure and lethal endpoints to long-term exposure and sublethal endpoints, and from laboratory exposures to field situations. Uncertainties associated with these extrapolations are incorporated into deterministic risk assessments by using assessment factors to either derive PNEC values from toxicity data or to set trigger values for TERs. The size of the assessment factor used to adjust toxicity data depends on the confidence with which the PNEC can be derived; the greater the confidence, the lower the assessment factor. For instance, when only short-term data are available, an assessment factor of 1000 is applied to the lowest L(E)C50, but if long-term toxicity data are available for species from three trophic levels (usually fish, *Daphnia*, algae), the assessment factor is reduced to 10. The assessment factor maybe as low as 1 if data are available from field or model ecosystem studies.[6] Similarly, trigger values for TER are set at 100 when based on short-term acute toxicity data, 10 when based on long-term toxicity data, and decided on a case-by-case basis when the TER is based on data from microcosm or mesocosm studies.[5]

Assessment factors used in deterministic risk assessments are empirically derived values that have limited scientific validity and whose use neither identifies nor quantifies sources of uncertainty. In contrast, probabilistic risk assessments enable estimates of uncertainty to be incorporated into both exposure and effects assessments. For example, species vary markedly in their sensitivity to environmental contaminants and uncertainties arising from interspecific variation can be described by constructing species sensitivity distributions (SSDs).[9] The SSD is estimated from single-species toxicity test data and visualised as a cumulative distribution function; it is used to calculate the concentration at which a specified fraction of the species pool will be affected by a chemical (*i.e.* hazardous concentration, HC). The most frequently estimated hazardous concentrations are the HC5 (5% of species affected) and HC10 (10% of species affected). Despite its widespread use in risk assessment, the SSD approach has been criticized for assuming that test species used to derive the SSD are random selections from the specified distribution and that they are representative of the ecosystem to be protected.[10] Recent studies with pesticides have demonstrated that hazardous concentrations derived from SSDs are protective of adverse ecological effects in freshwater systems and therefore do provide a useful tool for assessing environmental risk.[11,12] Probabilistic methods can also be applied to exposure estimates. They can incorporate spatial and temporal variation in the distribution and bioavailability of contaminants in the environment, as well as spatial and temporal variation in factors that influence the uptake of contaminants by the species of concern.[13] Probabilistic environmental risk assessments have recently been conducted for metals,[14] industrial chemicals[15] and pesticides.[16] In all cases, probabilistic risk assessments provided a higher degree of refinement than deterministic risk assessments and enabled management strategies to be targeted at those sites and species at greatest risk.

2 What are we Trying to Protect?

The purpose of environmental risk assessment is to inform the decision-making process. Therefore the first and most fundamental question that must be addressed when embarking on an environmental risk assessment is 'What are the protection goals?'. In other words, what are we trying to protect and over what temporal and spatial scales? For human risk assessment, the protection goal is the individual human being. However, the focus of an environmental risk assessment may range from the survival of individual members of an endangered species to the productivity of a community or the biodiversity of a region. Similarly, whereas it is relatively straightforward to identify the spatial and temporal scales of key determinants of the health and well-being of human populations, this is less obvious for ecosystems. For instance, whereas the appropriate spatial and temporal scales for microorganisms are measured in nanometers and seconds, for migratory fish or birds, they are measured in hundreds of kilometres and years.

Deciding what an environmental risk assessment is designed to protect requires input from stakeholders as well as scientists, and includes consideration of ecological values as well as ecological principles. Ecosystems provide society with many goods and services.[17] Essential ecosystem goods include the supply of oxygen, food, water, medicines and raw materials, and essential ecosystem services include nutrient cycling, water purification, waste removal, flood control, pollination and pest control.

Whereas it is relatively easy to value goods and services that are bought and sold (*e.g.* minerals, food, drinking water), it is much more difficult to value non-market goods and services such as biodiversity, nutrient cycling and climate regulation.[18] Despite these difficulties, the economic value of global ecosystem goods and services has been estimated to be about $US180 trillion based on 2000 prices;[19] a value which exceeds the Gross World Product by a factor of 4.5. Understanding how species interact to provide these goods and services and understanding how stressors, including chemicals, affect these interactions, is fundamental to ensuring the sustainable use of ecosystems.

Brock[20] identified three approaches for perceiving risk and setting protection goals; the pollution prevention principle, the carrying capacity principle and the functional redundancy principle. The pollution prevention principle adopts a precautionary approach by assuming that all environmental stressors are potentially harmful. The justification for adopting a conservative approach to risk assessment is the uncertainty with which the transport, fate, behaviour and effects of chemicals in the environment can be predicted. Environmental contaminants can move between media, be transported vast distances and rarely, if ever, occur in isolation.[21-23] Consequently, there is uncertainty in predicting which ecosystems will be exposed to which contaminants, at what concentrations and in what combinations. Furthermore, even if exposure could be predicted accurately and the direct impact of multiple stressors was known, there is uncertainty associated with predicting the long-term ecological consequences of those impacts. This includes understanding the indirect (*i.e.* trophic) effects of chemicals on species abundances and community structure[24] and the consequences of biodiversity loss for ecosystem functioning.[25]

The carrying capacity principle and the functional redundancy principle differ from the pollution prevention principle in that they assume that ecosystems can tolerate a certain degree of chemical stress. Where they differ from each other, is in the amount of stress and hence level of effect that is considered acceptable. The focus of the carrying capacity principle is the ability of ecosystems to resist and recovery from chemical perturbation (*i.e.* ecosystem resistance and resilience). Resistance is a measure of the capacity of a system to resist change, and resilience is either defined as the speed at which a system returns to equilibrium following a perturbation[26] or as the size of the disturbance from which a system cannot recover.[27] As sustainability of a system is dependent on its ability to withstand disturbances, it has been argued that resilience (*sensu* Holling) is an appropriate measure of sustainability[28] and hence is an ecologically relevant protection goal. The ability of an ecosystem to recover from chemical perturbation is currently considered as part of the European aquatic risk assessment of pesticides,[5] but it is not considered in the risk assessment of other chemicals.

The functional redundancy principle considers changes in community structure acceptable as long as key ecosystem functions are sustained. The application of this principle assumes that ecological roles can be performed by more than one species and therefore loss of species from a community does not necessarily result in loss of function. Whereas there is some evidence of functional redundancy within ecosystems, resulting in ecosystem processes being less sensitive to stressors than community structure,[29] this is not always the case,[30] and species that are 'redundant' under

Environmental Risk Assessment

one set of environmental conditions may not be redundant under another set of conditions. Consequently, there is an argument for maintaining high biodiversity even in communities with high functional redundancy, in order to maximise the chance that the system will be able to resist future stressors. The maintenance of high diversity to protect ecosystems against unpredictable and unknown stressors is known as the insurance hypothesis,[31] and has been likened to maintaining a diverse stock portfolio, that is, spreading risk by investing in many and varied companies. Moreover, there is increasing evidence to suggest that knowing which species are lost from a community is at least as important as knowing how many species are lost, when trying to predict the consequences of biodiversity loss on ecosystem function.[32]

The identification and adoption of different principles for setting protection goals, means that it is possible to establish an approach to risk assessment in which not all patches of habitat are treated equally. Areas with high ecological value (*e.g.* nature reserves, conservation areas) can be afforded a high degree of protection by applying the pollution prevention principle, whereas less pristine habitats in areas of intensive land use (*e.g.* agricultural land) could be adequately protected by adopting the carrying capacity or functional redundancy principle. The development of a targeted approach to protection goals setting and its potential application to environmental risk assessment is discussed by Brock *et al.*[33]

2.1 Assessment Endpoints

Once the protection goals have been established, the next task is to determine the most appropriate assessment endpoints. Ecosystems are comprised of many levels of organisation (*e.g.* individuals, populations, communities), multiple species and numerous processes and, except in the case of the protection of endangered species, it is rarely clear which species, process and level of organisation are the most critical for a risk assessment. Three criteria have been proposed to aid the selection of assessment endpoints: (1) ecological relevance; (2) susceptibility to known or potential stressors; and (3) relevance to protection goals.[34] Ecological relevance can be ascribed to endpoints at any level or organisation that help sustain the structure or function of the ecosystem under consideration. Ecologically relevant endpoints may relate to the provision of energy (*e.g.* primary production, decomposition) or habitat (*e.g.* vegetation structure) or to the structure of the community, ecosystem or landscape (*e.g.* species richness, habitat distribution). Susceptibility to environmental contaminants will depend on the mode of toxic action of the chemical, the exposure pathway and the life stage of the exposed organisms. Juvenile stages are often more sensitive than adults and vulnerability to stressors may be enhanced during physiologically demanding events such as migration, moulting and reproduction. Although ecological relevance and susceptibility are essential criteria for a risk assessment to be scientifically defensible, the ultimate aim of the risk assessment is to improve management decisions and therefore the assessment endpoints should include ecological components or attributes that people care about protecting. These include species and habitats with high conservation value (*e.g.* charismatic species) or important ecosystem goods and services (*e.g.* flood defence, waste disposal, food, timber, clean water, commercially or recreationally important species).

2.2 Impact and Causality

Once assessment endpoints have been measured the next questions to be addressed are:

1. is the system adversely affected (*i.e.* impacted)?
2. what is causing the impact?

These questions are relatively trivial for a single species exposed to a single chemical in a tightly controlled laboratory study, but they are less trivial when applied to natural ecosystems where multispecies assemblages are exposed to a variety of natural and anthropogenic stressors. Whereas the impact of a chemical on an ecosystem may be defined in terms of a deviation from a reference condition, a predefined change in an assessment endpoint, or a comparison with predicted values, the challenge with all these approaches is establishing the range of normal variation in the assessment endpoint so that extreme values can be identified. For instance, ecological quality is defined in the European Water Framework Directive in terms of deviation from 'undisturbed conditions'.[35] What is not clear, however, is what defines the undisturbed (reference) condition. Should it be defined in terms of present day conditions? If so, this raises the problem that in many European countries, including the UK, there are few if any sites that are not impacted by anthropogenic factors, therefore characterising a present-day baseline will be based on a limited number of potentially atypical sites. Should it be defined in terms of an historical state? If so, how far back in time is the baseline and what information do we use to characterise it? The definition of undisturbed (reference) condition is a non-trivial problem, as is the definition of ecological quality, yet both are fundamental to the successful implementation of the Water Framework Directive, which will apply to aquatic ecosystems across Europe.

A common approach to assessing causality is to compare 'contaminated' with 'uncontaminated' sites and to ascribed statistically significant differences in assessment endpoints to the contamination under investigation. The problem with this approach is that treatments are not randomly assigned to sites; sites are not true replicates and samples within sites are pseudoreplicated.[36] As a consequence, it is incorrect to infer that measured differences between 'contaminated' and 'uncontaminated' sites are due to the contaminant of interest. Causality can only be inferred by conducting experimental studies that demonstrate the relationship between cause and effect. Suter[2] proposed the following modification of Koch's postulates as a method to provide evidence of causality:

1. The injury, dysfunction or other putative effects of the toxicant must be regularly associated with exposure to the toxicant and any contributory causal factor.
2. Indicators of exposure to the toxicant must be found in the affected organisms.
3. The toxic effects must be seen when organisms or communities are exposed to the toxicant under controlled conditions, and any contributory factors should be manifested in the same way during controlled exposures.
4. The same indicators of exposure and effects must be identified in the controlled exposures as in the field.

Whereas the co-occurrence of the effect and contaminant in both the laboratory and field (postulates 1 and 3) is strong evidence for causation, proof is provided by demonstrating that the exposure conditions are associated with the same types and levels of effect in the field and laboratory (postulates 2 and 4).

3 What is the EU Legislative Framework?

Current EU chemicals legislation distinguishes between the 100,106 chemicals that were placed in the market before 1981 (*i.e.* 'existing' chemicals) and appoximately 2700 chemicals introduced since 1981 (*i.e.* 'new' chemicals). Whereas 'new' chemicals produced at volumes of 10 kg/year must be assessed for possible risks to the environment, there are no such provisions for 'existing' chemicals. An assessment of environmental risk is required only for those 'existing' chemicals that are identified as priority substances under Regulation 793/93 (*i.e.* about 140 chemical substances). In an attempt to rectify this anomaly, in February 2001, the European Commission issued a White Paper detailing its strategy for future chemicals policy.[37] The paper detailed a new system for ensuring a high level of chemical safety whilst maintaining a competitive chemicals industry, known as REACH (Registration, Evaluation, Authorisation, and Registration of Chemicals). REACH will apply to both 'new' and 'existing' chemical substances manufactured or imported in quantities of >1 tonne/year and will harmonise the environmental hazard and risk assessment of chemicals.

One of the major challenges posed by a harmonised scheme for 'new' and 'existing' chemicals is the backlog of 'existing' chemicals to be risk-assessed. Between 1993 and 2001, only 141 high-volume chemicals were identified for risk assessment, and only a limited number of these have completed the process.[38] Under REACH, all substances manufactured or imported into the European Union at quantities >1 tonne/year must be registered. Substances manufactured or imported in quantities >10 tonnes will require a hazard classification and an assessment of whether the substance is persistent, bioaccumulative and toxic (PBT) or very persistent and very bioaccumulative (vPvB). Exposure scenarios and risk management measures, which ensure that the risks from the uses of the substance are adequately controlled, must support substances classified as PBT or vPvB. Because of the large number of 'existing' chemicals requiring risk assessment, substances will be phased into REACH starting with the high production volume substances (>1000 tonnes/year) and those substances that are carcinogenic, mutagenic or toxic for reproduction.

Both representatives of the chemical industry and independent groups have suggested modifications to the REACH process. For instance, industry representatives have suggested a tiered approached based on existing EU risk assessment guidelines and tools (*i.e.* Technical Guidance Document, TGD[6]), which considers hazards and potential exposures simultaneously and which rapidly screens out chemicals and uses of no immediate concern.[39] The Royal Commission on Environmental Pollution have recommended that environmental monitoring is used to identify chemicals that require further investigation. Moreover, they suggest that synthetic chemicals that are found in biological material (*e.g.* breast milk) at elevated concentrations should be removed from the market, irrespective of whether they are known to cause harm.[40]

REACH will not apply to all chemicals and does not cover plant protection products, biocides, human medicines and veterinary medicines, which are regulated by different European directives. Assessment of the potential risks to the environment arising from veterinary medicines is required under Directive 2001/82/EC [41] and assessment of risks from human medicines is required under Directive 2001/83/EC.[42] Plant protection products (agricultural pesticides) are regulated under Directive 91/414/EEC,[43] whereas biocides (non-agricultural pesticides) are regulated under Directive 98/8/EC.[44] Biocides include products used as disinfectants or preservatives as well as for pest control (*e.g.* rodenticides, antifouling agents). Some form of environmental risk assessment is required for all chemicals. However, whereas the registration of most chemicals is dependent on the outcome of this risk assessment, this is not the case for human medicines. The environmental impact of human medicines must be assessed and arrangements made to limit them where appropriate, but environmental impact cannot be used to refuse marketing authorization in the European Union Member States (Directive 2004/27/EC).[38]

Environmental risk assessments for 'new' chemical substances, priority 'existing' chemical substances and biocidal products, are performed in accordance with the approaches detailed in the TGD. This includes guidance on how to determine PECs and PNECs, how to conduct a PBT assessment and how to decide on a testing strategy if further tests are needed. The purpose of the environmental risk assessment is to protect ecosystems and methodologies are specified for freshwater ecosystems, terrestrial ecosystems, marine ecosystems, air, top predators (secondary poisoning) and microorganisms in sewage treatment systems. The PEC can be derived using monitoring data or model calculations and the PNEC is usually determined using information from single-species studies, but could be based on information from model ecosystems. As the PNEC is regarded as the concentration below which an unacceptable effect is unlikely to occur, a PEC/PNEC >1 indicates unacceptable risk.

Assessment of the potential risks to the environment arising from medicinal products for human use is a step-wise procedure that emphasises the active substance and/or its metabolite(s) and which draws heavily on the TGD.[45] The first phase is a pre-screening stage (Phase I) in which environmental exposure is estimated. If the PEC for surface water is <0.01 $\mu g\ L^{-1}$, and no further environmental concerns are apparent, it is assumed that the product is unlikely to represent a risk to the environment following its prescribed usage by patients. If the PEC is greater than 0.01 $\mu g\ L^{-1}$ in surface water, then the product enters a screening stage (Phase II) in which standard aquatic toxicity and fate data are used to produce an initial assessment of risk based on PEC/PNEC ratios and PBT, vPvB criteria. In contrast to many other chemical substances, the initially risk assessment for human medicines is based on chronic toxicity data rather than acute toxicity data. This is because the specific mode of action of pharmaceuticals results in acute to chronic ratios being highly variable and often greater than 1000. Consequently, the assessment factor of 1000 specified in the TGD to be applied to acute data would not necessarily protect aquatic organism from chronic exposure to human medicines.[46] If the PEC/PNEC ratio exceeds 1, a refined risk assessment follows based on an extended dataset containing emissions, fate and effects information. This in turn may be followed by a site-specific risk assessment as described in the TGD.

A similar approached to that described for human medicines is employed to assess the environmental risk posed by veterinary medicinal products. However, in addition to environmental release via excretory products, the risk assessment for veterinary medicines also considers direct release of the product into the environment during its use (*e.g.* fish medicines, sheep dips, fumigation). The Phase I trigger values for veterinary medicines are an 'environmental introduction concentration' (*e.g.* concentration applied or in effluent) in water of 1 μg L^{-1} and a predicted environmental concentration in soil of 100 μg kg^{-1}; the exception being parasiticides administered to aquatic or pasture animals which, if enter the environment, automatically proceed to Phase II.[47] The Phase I environmental concentration is predicted using information on use (non-food animals and treatment of small flocks/herds are exempt), metabolism by treated animal (metabolised products are exempt), dosage and treatment regime, product type and manure production and disposal. If the substance proceeds to Phase II, it undergoes a two-tiered risk assessment, the precise form of which depends on whether the substance is used in aquaculture, administered to intensively reared terrestrial animals or administered to pasture animals.[48] For all uses, a standard set of physico-chemical and fate information is used to derive predicted environmental concentrations. In Tier A, acute toxicity data for algae, Crustacea and fish are used to derive PNEC data for aquatic systems and studies on earthworms, terrestrial plants and nitrogen transformation are used to derived PNEC values for terrestrial systems. In addition, if the substance is a parasiticide administered to pasture animals, a toxicity test using dung fauna should be conducted. If Tier A assessment results in a PEC/PNEC>1, then the substance moves on to Tier B, where information from chronic toxicity tests and fish bioconcentration studies are used to refine the risk assessment.[48] If the Tier B PEC/PNEC is still >1, then the risk assessment could possibly be refined further, by for example, testing metabolites or conducting microcosm or mesocosm studies.[49]

The most extensive risk assessment procedure is performed for plant protection products. A technical dossier on the active substance, formulation, and if appropriate metabolites, must be submitted before a plant protection product can be authorised for use in European Union Member States. The dossier contains information used to evaluate the risk that the product may pose to human health and the environment and includes, as a minimum, the following information: the identity, function, intended use and physical and chemical characteristics of the product; analysis of residues in soil, water, air, body fluids and tissues, food and feed; mammalian toxicity (acute toxicity, short-term toxicity, genotoxicity, reproductive toxicity, long-term toxicity and carcinogenicity) and metabolism; environmental fate and behaviour in soil, air and water; ecotoxicity to aquatic organisms and ecotoxicity to birds, non-target arthropods, microorganisms where appropriate. Pesticide risk assessment adopts a tiered approach with lower-tiers focusing on realistic worst-case exposure scenarios and acute toxicity and higher-tiers using information for more environmentally and ecologically realistic studies including mesocosm studies and field trials.[7] Environmental risk assessments may be based on deterministic or probabilistic approaches and studies used to support pesticide registration are performed in accordance with guidance documents produced by the European Commission.[5,50,51]

4 What are Some of the Challenges Associated with the Environmental Risk Assessment of Chemicals?

4.1 'Data-Poor' Chemicals

Approaches for assessing the environmental risks posed by plant protection products and 'new' chemicals are well established and risk characterisation is informed by robust datasets. However, these chemicals constitute a small fraction of the chemicals released into the environment and for most chemicals there is limited or little information on which to base the risk assessment. This 'data-poor' category includes most 'existing' chemicals, human and veterinary medicines, biocides and nanomaterials. A system has been proposed for addressing the lack of information for 'existing' chemicals (*i.e.* REACH), but there are challenges associated with implementing this system due to the large number of chemicals to be processed and the lack of appropriate techniques for detecting, quantifying and assessing some of the chemicals concerned. For many 'data-poor' chemicals not included in REACH, the challenges associated with assessing their environmental risk are even greater; this is particularly the case for nanomaterials and pharmaceuticals.

Nanomaterials have at least one dimension less than 100 nm and have been present in the environment for a long time. The adverse health effects of inhaling nanoparticles (*e.g.* ultrafine particles in polluted air) have been documented,[52] and there is increasing concern over the potential environmental risks posed by advances in nanotechnology. For example, there is concern that inhaled engineered nanomaterials will cause inflammatory reactions in the lungs similar to those resulting from the inhalation of coal dust and asbestos.[53] Engineered nanoparticles and nanotubes that are neither embedded nor fixed in a bulk material are free to be transported through environmental compartments and can enter organisms via ingestion, inhalation or movement through external surfaces (*e.g.* skin).[54] Moreover, once inside an organism, nanoparticles and nanotubes can move throughout the body and pass through biological membranes, including the blood–brain barrier.

A recent review of the potential health and environmental risks posed by nanotechnology[55] has recommended that the release of nanoparticles and nanotubes to the environment should be minimised until their risk can be properly evaluated. Moreover, it recommends that free nanoparticles are treated as new chemicals, even if they are comprised of materials that have already undergone a risk assessment. The rationale for treating nanomaterials as new chemicals is that changing the physical properties of materials also changes their environmental behaviour and toxicological properties.[56] The fact that nanoparticles have a greater potential to enter organisms than larger particles coupled with their toxicological potential[57] has raised the need for detailed studies of the environmental risks posed by these materials and has even led to the suggestion that a new subcategory of toxicology, namely nanotoxicology, should be defined to specifically address these knowledge gaps.[54]

Veterinary medicines can be released to the environment during manufacture, application or disposal and may be excreted by treated animals.[58] Evidence that veterinary medicines pose an environmental risk include the observation that many parasiticides are excreted in faeces in concentrations sufficient to disrupt the biology of dung fauna,[59] and the suggestion that population declines in vultures are caused by scavenging on the carcasses of treated animals.[60] The potential for the widespread use of veterinary medicines such as the parasiticide, ivermectin, to result in biodiversity loss and hence threaten the sustainability of pastoral ecosystems has been known for many years.[61] However, for a large proportion of veterinary medicines in current use, there is little or no information on which to base an assessment of the environmental risk of long-term, low-level exposure.[59] This information gap must be filled if the sustainability of agricultural systems is to be ensured.

Human and veterinary medicines contain bioactive substances that may elicit effects that are not detected in standard ecotoxicity studies; either because of inappropriate exposure durations, exposure concentration or response endpoints. The risk assessment of both human and veterinary medicines follows a two-phase process in which the first phase is an assessment of environmental exposure only. If veterinary medicines are introduced to the aquatic environment at concentrations <1 µg L^{-1} and human pharmaceuticals have a predicted environmental exposure to aquatic organisms of <0.01 µg L^{-1} (equivalent to an introductory concentration of 0.1 µg L^{-1}, assuming a 10-fold dilution), the chemical is assumed not to pose an environmental risk. Only if the environmental concentrations exceed these trigger values, does the assessment proceed to Phase II, in which both exposure and effects data are considered. These trigger values have been criticised for not being scientifically validated [62] and for not considering low-concentration effects (*e.g.* endocrine distruption) and mixture effects.[63] Furthermore, Phase II assessments have been criticized for their reliance on standard toxicity tests that do not take account of the fact that many pharmaceuticals have highly specialized modes of action.[64] The challenge, therefore, is to refine the risk assessment procedure to either remove or validate Phase I trigger values and to modify effects assessments to accommodate chemicals with very specific modes of action. Issues surrounding chemical mixtures and substances with non-classical modes of action are discussed below.

4.2 Non-Classical Modes of Action

Current environmental effects assessment procedures are designed to detect effects that occur during exposure and that can be measured in terms of changes in the survival, growth, reproduction or food consumption. They cannot be readily applied to substances with reversible, latent or trans-generational effects caused by exposure during a limited time window of sensitivity, nor to substances that result in effects not detected using standard ecotoxicological endpoints. Some chemicals may elicit effects that occur after exposure, either because the chemical is slow-acting, or because changes in the organism's physiology makes the chemical more toxic, or because the chemical disrupts a developmental process that only becomes apparent later in the life cycle of the organism. Latent effects have been reported for a number of carcinogens, endocrine disruptors and some pesticides.[65]

The inability of standard risk assessments to identify the environmental impacts of chemicals with non-classical modes of action has been highlighted by the discovery that chemicals may disrupt the endocrine system of organisms at very low exposure concentrations and that the consequences of endocrine disruption may not be manifest until a considerable time after exposure. Although chemicals may disrupt the endocrine system of biota by a wide variety of mechanisms, most research has focused on the effect of oestrogenic compounds, which include natural and synthetic (*i.e.* xeno-estrogens) substances that bind to the oestrogen receptor and elicit a feminising effect in organisms. A recent study of the contamination of Dutch waters by oestrogenic chemicals concluded that natural steroid hormones (*e.g.* estrone and 17ß-estradiol) were present in all untreated wastewaters and xeno-estrogens (*e.g.* bisphenol-A, alkylphenolethoxylates, phthalates) were present in almost all untreated wastewaters. Although the concentrations of most of these substances were lower in treated wastewater, estrone, bisphenol-A and several phthalates were each detected in around 50% of surface water samples.[66] Whereas discharges of municipal and industrial wastewater are the major routes of emissions of estrogenic compounds into the aquatic environment, in areas of intensive livestock farming, animal manure may also be an important source of environmental estrogens.[67]

Endocrine disruption is undoubtedly an issue that requires further investigation and is a mechanism of toxicity that must be picked up in the risk assessment procedure. However, there are potentially other toxic mechanisms that are not being detected and which may have important ecological consequences. The challenge is to improve our understanding of the toxic mode of action of chemicals and to use this understanding to develop more appropriate risk assessments.

4.3 Mixtures

Current regulatory risk assessments focus on individual substances acting on otherwise unstressed ecosystems. In reality, chemical contaminants rarely if ever, occur in isolation, and ecosystems are subject to a range of natural stressors. Two main concepts are used to predict the toxicity of mixtures: concentration addition and independent action. Concentration addition applies to mixtures of chemicals with the same mode of action and assumes that the effect of the mixture is the sum of the relative toxicities (expressed as toxic units) of the individual components. Moreover, concentration addition assumes that any chemical in the mixture can be substituted for another chemical with the same mode of action with the same relative toxicity, and that the overall mixture toxicity will remain the same. Independent action assumes that chemicals in mixtures have different modes of action and hence act independently. An important consequence of the difference between these two concepts is that, whereas with independent action chemicals in mixtures will only have an effect at concentrations above the NOEC, with concentration addition, chemicals can contribute to the total mixture effect even if they are present at concentrations below their individual NOEC.

Recent studies have demonstrated that the mixture toxicity of anti-inflammatory drugs and the mixture toxicity of β-blockers can be accurately predicted using the concept of concentration addition.[68, 69] One implication of this finding is that mixtures of pharmaceuticals may have environmental effects even when the individual chemicals are present at concentrations below the Phase I trigger level. Other studies have also demonstrated that mixtures containing chemicals at concentrations considerably below NOEC values (*e.g.* EC1) exhibit considerable toxicity.[70] These research findings highlight the need to improve our understanding of mixture toxicity, especially when dealing with chemicals present at low-effect concentrations, and to modify environmental risk assessment procedures to take account of possible combined effects resulting from multiple chemical exposures.

References

1. C.J. Van Leeuwen and J. Hermans, *Risk Assessment of Chemicals: An Introduction*, Kluwer Academic Publishers, Dortrecht, The Netherlands, 1995.
2. G.W. Suter II, *Ecological Risk Assessment*, Lewis Publishers, Boca Raton. FL, 1993.
3. D. Mackay, E. Webster and T. Gouin, *Issues Environ. Sci. Technol.*, **22**, chapter 7, this volume.
4. P. Harrison and P. Holmes, *Issues Environ. Sci. Technol.*, **22**, chapter 4, this volume.
5. EC, *Guidance Document on Aquatic Toxicology in the Context of Directive 91/414/EEC. SANCO/3268/2001*, Brussels, Belgium, 2002.
6. EC, *Technical Guidance Document on Risk Assessment*, European Commission, Brussels, 2003.
7. P.J. Campbell, D.J.S. Arnold, T.C.M. Brock, N.J. Grandy, W. Heger, F. Heimbach, S.J. Maund and M. Streloke, *Guidance Document on Higher-Tier Aquatic Risk Assessment for Pesticides (HARAP)*, SETAC-Europe, Brussels, 1999.
8. K. Solomon, J. Giesy and P. Jones, *Crop Prot.*, 2000, **19**, 649.
9. L. Posthuma, G.W. Suter II and T.P. Traas, *Species Sensitivity Distributions in Ecotoxicology*, Lewis Publishers, Boca Raton, FL, 2002.
10. V. Forbes and P. Calow, *Hum. Ecol. Risk Assess.*, 2002, **8**, 473.
11. L. Maltby, N. Blake, T.C.M. Brock and P.J. Van den Brink, *Environ. Toxicol. Chem.*, 2005, **24**, 379.
12. P.J. Van den Brink, N. Blake, T.C.M. Brock and L. Maltby, *Hum. Ecol. Risk Assess.*, in press.
13. A. Hart, G.C. Smith, R. Macarthur and M. Rose, *Toxicol. Lett.*, 2003, **140–141**, 437.
14. P.A. Van Sprang, F.A.M. Verdonck, P.A. Vanrolleghem, M.L. Vangheluwe and C.R. Janssen, *Environ. Toxicol. Chem.*, 2004, **23**, 2993.
15. M. Zolezzi, C. Cattaneo and J.V. Tarazona, *Environ. Sci. Technol.*, 2005, **39**, 2920.
16. J.J. Johnston, W.C. Pitt, R.T. Sugihara, J.D. Eisemann, T.M. Primus, M.J. Holmes, J. Crocker and A. Hart, *Environ. Toxicol. Chem.*, 2005, **24**, 1557.
17. R. Costanza, R. d'Arge, R. de Grrot, S. Farber, M. Grasso, B. Hannon, K. Limburg, S. Maeem, R.V. O'Neill, J. Paruelo, R.G. Raskin, P. Sutton and M. Van den Belt, *Nature*, 1997, **387**, 253.
18. M.A. Wilson and S.R. Carpenter, *Ecol. Appl.*, 1999, **9**, 772.
19. R. Boumans, R. Costanza, J. Farley, M.A. Wilson, R. Portela, J. Rotmans, F. Villa and M. Grasso, *Ecol. Econ.*, 2002, **41**, 529.
20. T.C.M. Brock, in *Workshop on Risk Assessment and Risk Mitigation Measures in the Context of the Authorization of Plant Protection Products (WORMM)*, Vol. 383, R. Forster and M. Streloke (Eds.), Mitteilungen aus der Biologischen Bundesanstalt für Land- und Forstwirtschaft, Berlin-Dahlem, 2001, 68–72.

21. F. Wania, *Environ. Sci. Technol.*, 2003, **37**, 1344.
22. D.C.G. Muir, C. Teixeira and F. Wania, *Environ. Toxicol. Chem.*, 2004, **23**, 2421.
23. K. Vorkamp, F. Riget, M. Glasius, M. Pécseli, M. Lebeuf and M. D, *Sci. Total Environ.*, 2004, **331**, 157.
24. J.W. Fleeger, K.R. Carman and R.M. Nisbet, *Sci. Total Environ.*, 2003, **317**, 207.
25. M. Loreau, S. Naeem and P. Inchausti, *Biodiversity and Ecosystem Functioning*, Oxford University Press, Oxford., 2002.
26. S.L. Pimm, *Nature*, 1984, **307**, 321.
27. C.S. Holling, *Ann. Rev. Ecol. Syst.*, 1973, **4**, 1.
28. M.S. Common and C. Perrings, *Ecol. Econ.*, 1992, **6**, 7.
29. D.M.E. Slijkerman, D.J. Baird, A. Conrad, R.G. Jak and N.M. van Straalen, *Environ. Toxicol. Chem.*, 2004, **23**, 455.
30. D.M. Carlisle and W.H. Clements, *Freshwat. Biol.*, 2005, **50**, 380.
31. S. Yachi and M. Loreau, *PNAS*, 1999, **96**, 1463.
32. D.U. Hooper, F.S. Chapin III, J.J. Ewel, A. Hector, P. Inchausti, J. Lawton, D.M. Lodge, M. Loreau, S. Naeem, B. Schmid, H. Setälä, A.J. Symstad, J. Vandermeer and D.A. Wardle, *Ecol. Monogr.*, 2005, **75**, 3.
33. T.C.M. Brock, G.H.P. Artes, L. Maltby and P.J. van den Brink, *Aquatic Hazards of Pesticides, Ecological Protection Goals and Common Claims in EU Regulation: An Attempt to Harmonise Scientific Methods to Derive Permissible Concentrations*, Alterra Report, Wageningen, The Netherlands, in press.
34. EPA, *Guidelines for Ecological Risk Assessment. EPA/630/R-95/002F*, US Environmental Protection Agency, Washington, DC, 1998.
35. EC, *Official J. Euro. Commun.*, 2000, **L 327**, 1.
36. S.H. Hurlbert, *Ecol. Monogr.*, 1984, **54**, 187.
37. CEC, *Strategy for a Future Chemicals Policy COM(2001) 88 final*, Commission of the European Communities, Brussels, 2001.
38. EC, *REACH in Brief*, European Commission, Brussels, Belgium, 2004.
39. T. Feijtel, M. Comber, W. De Wolf, M. Holt, V. Koch, A. Lecloux and A. Siebel-Sauer, *Environ. Toxicol. Chem.*, 2005, **24**, 251.
40. RCEP, *Chemicals in Products–Safeguarding the Environment and Human Health*, HMSO, London, 2003.
41. EC, *Official J. Euro. Commun.*, 2001, **L311**, 1.
42. EC, *Official J. Euro. Commun.*, 2001, **L331**, 67.
43. EC, *Official J. Euro. Commun.*, 1991, **L230**, 1.
44. EC, *Official J. Euro. Commun.*, 1998, **L123**, 1.
45. EMEA, *Guideline on the Environmental Risk Assessment of Medicinal Products for Human Use (CHMP/SWP/4447/00 draft)*, European Medicines Agency, London, 2005.
46. B. Ferrari, R. Mons, B. Volleat, B. Fraysse, N. Paxéus, R. Lo Giudice, A. Pollio and J. Garric, *Environ. Toxicol. Chem.*, 2004, **23**, 1344.
47. VICH, *Environmental Impact Assessment (EIAs) for Veterinary Medicinal Products (VMPs) – Phase I. VICH GL6 (Ecotoxicity Phase I)*, International Cooperation on Harmonisation of Technical Requirements for Regulation of Veterinary Medicinal Products, Brussels, Belgium, 2000.

48. VICH, *Environmental Impact Assessment for Veterinary Medicinal Products Phase II Guidance. VICH GL38 (Ecotoxicity Phase II)*, International Cooperation on Harmonisation of Technical Requirements for Regulation of Veterinary Medicinal Products, Brussels, Belgium, 2004.
49. P.J. Van den Brink, J.V. Tarazona, K.R. Solomon, T. Knacker, N.W. Van den Brink, T.C.M. Brock and J.P. Hoogland, *Environ. Toxicol. Chem.*, 2005, **24**, 820.
50. EC, *Guidance Document for Risk Assessment for Birds and Mammals Under Council Directive 91/414/EEC. SANCO/4145/2000*, Brussels, Belgium, 2002.
51. EC, *Guidance Document on Terrestrial Ecotoxicology Under Council Directive 91/414/EEC. SANCO/10329/2002*, Brussels, Belgium, 2002.
52. G. Oberdorster, *Int. Arch. Occup. Environ. Health*, 2001, **74**, 1.
53. R. Owen and M. Depledge, *Mar. Pollut. Bull.*, 2005, **50**, 609.
54. K. Donaldson, V. Stone, C.L. Tran, W. Kreyling and P.J.A. Borm, *Occup. Environ. Med.*, 2004, **61**, 727.
55. Royal Society and Royal Academy of Engineering, *Nanoscience and Nanotechnologies: Opportunities and Uncertainties*, www.nanotec.org.uk/final Report.htm, 2004.
56. K. Donaldson, V. Stone, P.S. Gilmour, D.M. Brown and W. MacNee, *Phil. Trans. R. Soc. Lond. Ser, A*, 2000, **358**, 2741.
57. V.L. Colvin, *Nat. Biotechnol.*, 2003, **21**, 1166.
58. A.B.A. Boxall, D.W. Kolpin, B. Halling-Sørensen and J. Tolls, *Environ. Sci. Technol.*, 2003, **37**, 286A.
59. K.G. Wardhaugh, *Environ. Toxicol. Chem.*, 2005, **24**, 789.
60. J.L. Oaks, M. Gilbert, M.Z. Virani, R.T. Watson, C.U. Meteyer, B.A. Rideout, H.L. Shivaprasad, S. Ahmed, M.J.I. Chaudhry, M. Arshad, S. Mahmood, A. Ali and A.A. Khan, *Nature*, 2004, **427**, 630.
61. R. Wall and L. Strong, *Nature*, 1987, **327**, 418.
62. CSTEE, *Opinion on Draft CPMO Discussion Paper on Environmental Risk Assessment of Medicinal Products for Human Use [Non-Genetically Modified Organism (Non-GMO) Containing]*, Scientific Committee on Toxicology, Ecotoxicology and the Environment, Brussels, Begium, 2001.
63. J.P. Bound and N. Voulvoulis, *Chemosphere*, 2004, **56**, 1143.
64. K.-P. Henschel, A. Wenzel, M. Diedrich and A. Fliedner, *Reg. Toxicol. Pharm.*, 1997, **25**, 220.
65. M.K. Hurd, S.A. Perry and W.B. Perry, *Environ. Toxicol. Chem.*, 1996, **15**, 1344.
66. A.D. Vethaak, J. Lahr, S.M. Schrap, A.C. Belfroid, G.B.J. Rijs, A. Gerritsen, J. de Boer, A.S. Bulder, G.C.M. Grinwis, R.V. Kuiper, J. Legler, T.A.J. Murk, W. Peijnenberg, H.J.M. Verhaar and P. de Voogt, *Chemosphere*, 2005, **59**, 511.
67. T.A. Hanselman, D.A. Graetz and C. Wilkie, *Environ. Sci. Technol.*, 2003, **37**, 5471.
68. M. Cleuvers, *Ecotoxicol. Eviron. Safety*, 2004, **59**, 309.
69. M. Cleuvers, *Chemosphere*, 2005, **59**, 199.
70. T. Backhaus, R. Altenburger, A. Arrhenius, H. Blank, M. Faust, A. Finizio, P. Gramatica, M. Grote, M. Junghans, W. Meyer, M. Pavan, T. Porsbring, M. Scholze, R. Todeschini, M. Vighi, H. Walter and L. Horst Grimme, *Continental Shelf Res.*, 2003, **23**, 1757.

Risk Assessment of Metals in the Environment*

PETER G.C. CAMPBELL, PETER M. CHAPMAN AND BEVERLEY A. HALE

1 Introduction

The field of ecological risk assessment (ERA) was developed in the 1980s, in recognition of the need to evaluate the effects of human activities at the ecosystem level and to incorporate this ecological information into regulatory decisions. ERA can be defined as "A process that evaluates the likelihood that adverse ecological effects are occurring or may occur as a result of exposure to one or more stressors".[1] Environmental contamination by synthetic organic molecules constitutes one of the more obvious potential stressors, and indeed many of the initial applications of ERA dealt with this class of environmental contaminants. As a result, the initial methods used to assess the risk posed by environmental contamination, and the criteria used to screen contaminants as a function of their potential hazard to the environment, were developed with a focus on synthetic organic molecules. Subsequent attempts to apply this methodology to inorganic contaminants proved unsuccessful,[2] much to the dismay of environmental regulators who understandably would have preferred to use a uniform methodology and a single set of criteria. Recent research has confirmed the limitations of the traditional ERA approach, when applied to metals and metalloids, and considerable effort has been expended to identify possible adaptations to the methodology that would make it more amenable to this ubiquitous class of contaminants.[3,4]

In this paper we review the basic principles of ERA, compare and contrast organic and inorganic (metallic) contaminants, identify where the traditional ERA methodology fails for metals, and finally suggest possible modifications to current ERA procedures. Note that we have used the term "metals" in a very general sense. Our examples

---

* Contribution No. 48 from the Metals in the Environment Research Network (MITE-RN: www.mite-rn.org)

focus on those electropositive elements that exist to the left of the periodic table; many of these elements exist as cations or their complexes in the natural environment (*e.g.*, Ag, Cd, Cu, Hg, Pb, Zn), but others are present as oxyanions (*e.g.*, Mo, V). In addition, we also consider some of the metalloids (*e.g.*, As, Sb), *i.e.* those elements found along the stair-step line that distinguishes metals from non-metals in the periodic table, drawn from the border between B and Al to the border between Po and At.

2 Ecological Risk Assessment

2.1 Historical Overview

Risk assessment includes a framework for gathering data and evaluating both their sufficiency and their certainty relative to decision-making. Typically, risk assessment involves two components: an assessment of risk, followed by attempts to minimize unacceptable risks based on that assessment. These two components have been formalized in risk assessment into, respectively, the scientific process of human and ecological risk assessment and the separate process of risk management.

Human risk assessment has always been personal and individual, ranging from prehistoric hunting parties to modern humans crossing busy streets. The realization that industrial chemicals could be hazardous outside the laboratory or factory resulted in the development of human health risk assessment (HHRA); the U.S. Environmental Protection Agency (U.S. EPA) began developing guidelines for HHRA in the 1970s. Development of similar guidelines for ecological effects began in the 1980s, and the U.S. EPA published their framework for ERA in 1992.[1] The European Union (EU) subsequently published a Technical Guidance Document on risk assessment of new or existing substances in 1996.[5]

There are key differences between HHRA and ERA. Whereas the former is concerned with only one species and with protection of all individuals of that species, the latter is concerned with many species. Except for threatened or endangered species, the goal of ERA is the protection not of individuals but rather of populations and communities. This may take the form of protecting the most sensitive species. Alternatively, some level of damage to individuals may be allowed, provided that populations and communities are not harmed (*e.g.*, the goal in some jurisdictions for the protection of aquatic species is to protect 95% of species 95% of the time).

Originally, ERA of necessity focused on heavily contaminated areas where the contaminants could, and in many cases did, cause adverse effects on biota living (or attempting to live) in those areas. However as such areas were assessed and, where necessary, remediated (at least in developed countries), it was realized that effects of contaminants are not necessarily restricted to localized areas, nor are chemicals (organic or inorganic substances) the only stressors of potential concern. Stressors can also be physical (*e.g.*, habitat loss) or biological (*e.g.*, introduced species), and combinations of different stressors can increase risks beyond localized sites. Thus, ERA is now also conducted at the regional or watershed level and considers non-chemical stressors.[6]

In addition, ERA now extends beyond the original practice of simply retroactively assessing risk. The ERA approach is now also used to predict risk, to compare the risk to the environment posed by different actions or developments, to rank risks posed by different stressors and to develop site-specific guidelines as necessary and appropriate. In all of these applications, related to chemical contaminants, ERA typically focuses on the question "How much contamination is 'too much'?"; in other words, "What are 'safe' levels of contamination?"

2.2 Basic Approach

Although ERA begins by evaluating contamination (substances present where they should not be, or present in excess of baseline levels), its focus is not on contamination *per se* but rather on pollution (contamination resulting in adverse biological effects: all pollutants are contaminants, but not all contaminants are pollutants). Similarly, the initial evaluation is of hazard (the possibility of an effect), while the final evaluation is of risk (the probability of an effect). An ecological risk does not exist unless two conditions are satisfied: (1) the stressor has the inherent ability to cause one or more adverse effects and (2) the stressor co-occurs with or contacts an ecological component with sufficient duration and magnitude to elicit the identified adverse effect. Thus, the ERA for a specific contaminant is based on the determination and comparison of the predicted (or measured) environmental concentration (PEC) of the contaminant and its predicted (or measured) no effect concentration (PNEC), the PEC/PNEC ratio being referred to as the hazard quotient (HQ). If the PEC and/or PNEC are based on generic rather than site-specific information, then the PEC/PNEC comparison only provides an indication of hazard. In such a case, if PEC/PNEC<1, there is negligible risk; if the ratio is >1, there is the possibility but not the certainty of risk. Determining whether risk is probable requires site- and contaminant-specific information, as described for metals in Section 4. Such information can be used in simple comparisons as for the hazard quotient described above, or in more complex probabilistic comparisons.

The ERA frameworks used in different jurisdictions tend to be more similar than different. In the USA, the ERA framework has three components: hazard identification, exposure and effects characterization, and risk characterization. In Canada, the EU and Australia there are four components, with some differences in nomenclature: problem formulation, exposure assessment, effects (or dose-response) assessment, and risk characterization. Other countries such as Japan, China and Mexico do not have formal ERA guidance in place and use frameworks developed elsewhere. Key differences among countries include the levels of protection required; for example, the USA does not require protection of the most sensitive species, but Canada does.

3 Comparison of Inorganic and Organic Contaminants

As alluded to earlier, the application of standard ERA methods to metals often leads to illogical results and impractical conclusions.[2] From an environmental chemistry/toxicology perspective, this failure is hardly surprising. The factors controlling the fate and effects of metals in the environment differ from those controlling the fate and effects of other "contaminants" present in the environment, such as synthetic

Table 1 Comparison of metals and synthetic organic contaminants

Characteristic	Synthetic organic contaminants	Metals
Origin	Anthropogenic; background concentrations → 0	Geogenic (naturally occurring); background concentrations variable
Fate	Subject to various degradation processes (hydrolysis, photolysis, biodegradation); environmental half-lives meaningful; parent → daughter compounds (irreversible changes); biomagnification common	Not subject to degradation processes; infinitely persistent; environmental half-lives not meaningful; changes in metal speciation (reversible); biomagnification rare (exception: methyl-Hg)
Interactions with living cells	Uptake often by simple diffusion across a bilipid barrier; generic narcotic effects	Uptake normally by facilitated transport; metal-specific toxicity
Biological function	None; bioaccumulation not a natural phenomenon	Often biologically essential; bioaccumulation a natural phenomenon

organic compounds. In this section, we briefly summarize the principal differences between these two classes of contaminants (Table 1).

3.1 Origin and Environmental Fate

Metals occur in the environment at concentrations ranging from the geological background to potentially much higher levels in areas affected by natural point sources or by human activities. Whereas background concentrations of synthetic organic molecules can safely be assumed to be negligible, for metals one must consider the (variable) geological background concentrations and their long-term influence on the indigenous fauna and flora.

Organic contaminants are subject to various degradation processes, both abiotic (hydrolysis, photolysis) and biotic (biodegradation).[7] These processes generally involve changes in covalent bonding and are normally irreversible, leading from the parent compound to various daughter molecules; one can determine meaningful half-lives for the parent compound and its degradation products in various environmental compartments. Metals on the other hand are inherently persistent; they are neither created nor destroyed by anthropogenic or biological processes, although they may undergo various speciation changes.

Speciation refers to distribution of a metal among various chemical species, where "species" refers to a specific form of an element defined as to isotopic composition, electronic or oxidation state and/or complex or molecular structure[8]; examples of environmentally relevant metal species are shown in Table 2. Transformations among various metal species often involve changes in (weak) coordinate bonding, or changes in oxidation state; unlike the case for organic molecules, these changes are normally reversible, with the important consequence that the speciation of a metal is a function of the chemistry of the medium in which it is found. In contrast, the "speciation" of

Table 2 *Metal speciation – examples of metal forms found in surface waters*

Species	Examples
Free metal ions	$Al^{3+}(H_2O)_6$
	$Cu^{2+}(H_2O)_6$
Hydroxo-complexes	$AlOH^{2+}$, $Al(OH)_2^+$, $Al(OH)_4^-$
	$FeOH^{2+}$, $Fe(OH)_2^+$, $Fe(OH)_4^-$
	$Cu(OH)_2^0$
Simple inorganic complexes	AlF^{2+}, AlF_2^+
	$CdCl^+$, $CdCl_2^0$, $CdCl_3^-$
	$HgCl_2$, $HgOHCl^0$
	$CuCO_3^0$
	$CdSO_4^0$
Simple organic complexes	
synthetic	Cu-$EDTA^{2-}$
	Cd-NTA^{1-}
natural	Cd-alanine
	Cd-citrate
	Fe-siderophore
Polymeric organic complexes	Al, Fe, Cu, Pb or Hg – fulvic or humic acids

an organic contaminant is largely determined by the form in which the contaminant enters the environment and by any subsequent degradation processes.

3.2 Biological Interactions

Biological uptake of most synthetic organic contaminants occurs by simple passive diffusion across a cell membrane. Accordingly, the octanol–water partition coefficient of the contaminant (K_{ow}) is often a good predictor of its tendency to bioaccumulate.[7] Membrane carriers are not involved, and the biological effect of the contaminant is often characterized by narcosis. In contrast, metals generally exist in the environment in polar, hydrophilic forms (Table 2), which are strongly hydrated and are unable to traverse biological membranes by simple diffusion. Membrane transport occurs by facilitated transport, usually passive (*i.e.*, not against a concentration gradient), and necessarily involving either membrane carriers or channels[9]; with few exceptions (*e.g.*, some organometallics and neutral metal complexes, such as CH_3HgCl and $HgCl_2^0$, respectively), the octanol-water partitioning coefficient of the metal forms has no bearing on the relative facility with which they cross biological membranes. Similarly, a variety of mechanisms exists for the detoxification of metals once they have entered a living cell,[10] precluding simple predictions of metal-induced toxicity on the basis of metal quotas or burdens, as is done with organic contaminants.

The fact that the Earth's fauna and flora have evolved in the presence of metals has important repercussions from an environmental risk perspective, notably with respect to such questions as essentiality, bioaccumulation and tolerance (acclimation and adaptation). Many metals and metalloids are required for biological life (*e.g.*, Cu, Co, Fe, Mn, Mo, Ni, Se, and Zn),[2] and therefore their presence in the ambient environment is essential; clearly, synthetic organic contaminants do not fall into this category. Bioaccumulation of essential metals is a natural process required by living organisms for metabolism and growth; many organisms have developed effective means of regulating internal essential metal concentrations within quite narrow ranges (homeostasis). These homeostatic control mechanisms can also play a role in the detoxification of non-essential metals such as Pb or Hg.

Given these obvious differences between organic and inorganic contaminants, it should not be surprising that ERA methods developed for the former fail when applied without discernment to metals and metalloids. In the following section we identify the major problems with the generic ERA approach, as applied to the aquatic and terrestrial environments.

4 Problems with the Application of Traditional ERA Approaches to Metals

4.1 Generic Considerations

4.1.1 Persistence–Bioaccumulation–Toxicity

The realization that synthetic organic chemicals could be hazardous to human health and the environment originated with certain organic chemicals that caused adverse effects to humans and charismatic wildlife (*e.g.*, eagles) via biomagnification – they increased in concentration through three or more trophic levels of the food chain, resulting in adverse effects to top predators. These chemicals included dichloro-diphenyl-trichloroethane (DDT), polychlorinated biphenyls (PCBs) and tetrachloro-dibenzo-*p*-dioxin (2,3,7,8-TCDD); adverse effects included egg shell thinning in eagles and other raptors.

The hazard and risk posed by these chemicals could be estimated on the basis of their persistence, bioaccumulation and toxicity (PBT).[11] Specifically, the less persistent a compound was, and the lower its tendency to accumulate in biological tissues (bioaccumulate), the less problematic it was. Because the PBT approach worked and was relatively simple, it was generically applied to categorize the potential risk of other contaminants. However, as discussed in Section 3, contaminants do not all behave like hydrophobic organic molecules! Unlike man-made chemical substances such as DDT, and with the exception of specific radioisotopes, metals can neither be created nor destroyed. Accordingly, the traditional "persistence" criterion does not yield a meaningful hazard or risk ranking for metals. In recognition of this dilemma, and in an effort to develop a consistent approach for evaluating the persistence of organic and inorganic substances, it has recently been suggested that persistence for both classes of contaminants could be assessed with a common methodology in which mass balance models are used to determine the substance's behaviour in a specified environment or "unit world".[3,12]

The "bioaccumulation" criterion for metals is similarly ineffective. The bioaccumulation potential of an organic contaminant is typically assessed by determining either a BAF or a BCF. A BCF is the ratio of the concentration of substance in an organism to the same substance in the surrounding environment (water) from which it was taken up; a BAF not only considers the waterborne contaminant but also accounts for uptake from food. This approach works well for organic substances, where uptake is based on simple diffusion and lipid partitioning; for environmentally realistic contaminant concentrations, uptake is a linear function of the exposure concentration and accordingly

BCFs and BAFs are reasonably constant from one environment to another. This approach fails for metals, however, since uptake occurs by facilitated transport and typically exhibits saturation (Michaelis–Menten) kinetics.[9] Furthermore, some metals such as zinc and copper are essential and their bioaccumulation, unlike that of manmade chemicals, is required for life cycle completion for all organisms. In such cases, organisms will normally attempt to maintain relatively constant internal metal concentrations: if ambient metal concentrations are low, the BCFs and BAFs will thus tend to *increase* (a constant numerator divided by a small denominator); conversely, if ambient metal levels are high, the BCFs and BAFs will *decrease* (Figure 1).[13] Clearly, the BCFs and BAFs are not an intrinsic property of the metal, but rather integrate the nature of the metal and the ambient exposure conditions. Note, too, that the rationale behind the BCF/BAF approach, *i.e.*, the need to account for the biomagnification of (organic) contaminants through three or more trophic levels of the food chain, does not apply to (inorganic) metals. Mercury is the only metal that biomagnifies, and it only does so in its methylated (organic) form. It follows that BCF or BAF values for metals bear no relationship to potential environmental risk.[3] Secondary poisoning due to metal concentration increases through two trophic levels of the food chain is, however, a possibility,[14] as described below.

To summarize, given the pervasive influence of metal homeostasis and metal detoxification (see below), the bioaccumulation of a metal is rarely sufficient to predict environmental risk. It follows that neither BAFs nor BCFs will be useful for ERA of metals.[3,15]

4.1.2 Bioavailability – External Environment
The bioavailability of organic contaminants is generally considered to be an intrinsic property of the original parent molecule, although such environmental factors as the concentration of natural dissolved organic matter (DOM) can modulate this bioavailability to some extent.[16] In contrast, environmental factors play a much more important role in determining the bioavailability of metals, in large part because the bioavailability of a given metal is so strongly influenced by its speciation, which in turn is largely determined by the chemical conditions in the receiving environment (*e.g.*, pH, water hardness, salinity, concentrations of dissolved organic matter in surface waters, sediment pore water, soil pore water, and ground water). These same chemical conditions can also exert a direct effect on the exposed organism and affect its sensitivity to the metal.

Given the marked influence of speciation on metal bioavailability, there is a clear need within ERA for chemical equilibrium models that can be used to predict metal speciation in the receiving environment, as a function of pH, ionic strength and ligand concentrations. Many such models exist, and for inorganic metal species (*e.g.*, hydroxo-, chloro-, sulphato- and carbonato-complexes) in reasonably dilute systems they yield accurate and useful predictions. However, natural systems normally include variable amounts of fulvic and humic acids, which cannot be modelled successfully as simple, low-molecular weight, mono-dentate ligands.[17] Two advanced models have been developed recently to take into account the poly-electrolytic nature of fulvic and humic acids, and their poly-functionality: the Windermere humic aqueous model (WHAM)[18] and the nonideal competitive adsorption-Donnan (NICA-Donnan) model.[19,20] These models

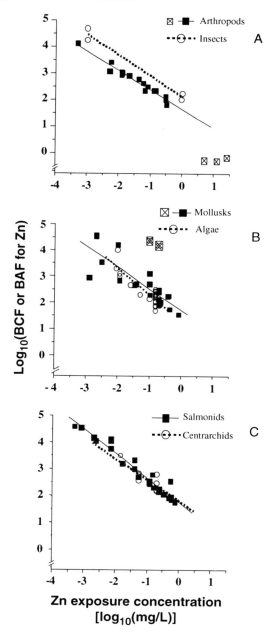

Figure 1 *Effect of chronic Zn exposure on BCF and BAF values in aquatic biota. (A) arthropods and insects; (B) mollusks and algae; (C) salmonids and centrarchids. Data are presented on a log-log basis and the best-fit line from the linear regression analysis is shown. Adapted from McGeer et al.[13] With permission from Environmental Toxicology and Chemistry, Alliance Communications Group*

have been calibrated on the basis of laboratory titrations of isolated fulvic and humic acids, and they are able to reproduce these titration curves quite successfully.[17] What is less obvious, however, is whether they can also predict metal complexation in natural waters, where the total metal levels are often at or below the lowest levels used in the laboratory titrations, where the proportions of fulvic and humic acid vary spatially and temporally, where Fe(III), Al and other metals may compete for available binding sites[21] and where the organic matter is in its natural state (*i.e.,* it has not been subjected to the rigours of a laboratory isolation and purification procedure). With respect to this latter point, it has been suggested that the commonly used isolation and purification protocols may irreversibly alter the chemical nature of the natural organic ligands (*e.g.,* through oxidation of reduced sulfur species associated with the fulvic and humic acids).[22] There is a real need for measurements of the degree of metal complexation in minimally perturbed natural systems, and for comparisons between these measurements and the predictions of the equilibrium complexation models.

4.1.3 Bioavailability – Internal Environment

Bioavailability of metals also varies within organisms; the term "bioreactive metals" has been used to describe metals within tissues that are not biologically detoxified or inert.[15] Metals bound to inducible metal-binding proteins such as metallothionein (MT), or precipitated into insoluble concretions consisting of metal-rich granules, can be considered to be biologically detoxified metal (BDM), as compared to metals in metal-sensitive fractions (MSF) such as organelles and heat-sensitive proteins.[23,24] The relative proportions of BDM and MSF vary among species, a higher proportion of MSF presumably resulting in greater metal vulnerability for that organism. This distinction between BDM and MSF builds on earlier work differentiating between two pools of metals in organisms – one that is metabolically active and available, and the other that has been detoxified and is unavailable[10,25]; this latter pool can have virtually unlimited potential for metal accumulation, at least for marine invertebrates.

A corollary to this model of metal accumulation is that metal tolerance or resistance will be related to the ability of an organism to prevent "inappropriate" metals from binding to sensitive sites. The binding of an "inappropriate" metal to a metal-sensitive site, often termed "spillover", could be the precursor to the onset of metal-induced stress.[10] A limited number of laboratory exposures have been used to test this spillover model of metal toxicity, as have field experiments involving transplantation or habitat swap designs. The typical experimental design involves subjecting the test organism to a change in metal exposure, following metal accumulation in various sub-cellular fractions over time, and noting the onset of toxicity. In such experiments, there is some evidence linking the onset of metal toxicity to the appearance of metals in putative metal-sensitive pools.[26–28] However, the situation with chronically exposed organisms is less clear. Along a metal exposure gradient, one might expect that the indigenous organisms should be able to detoxify the metal effectively up to a certain exposure threshold, *i.e.,* the contribution of MSF to the total metal body burden should be low. Beyond this exposure threshold, metal spillover into the MSF should occur and metal-induced effects should appear (*i.e.,* a threshold or "hockey-stick" response on proceeding from site to site along the metal concentration gradient – see Figure 2). Very few data are available to test this model, but in a recent study of indigenous aquatic

Risk Assessment of Metals in the Environment   111

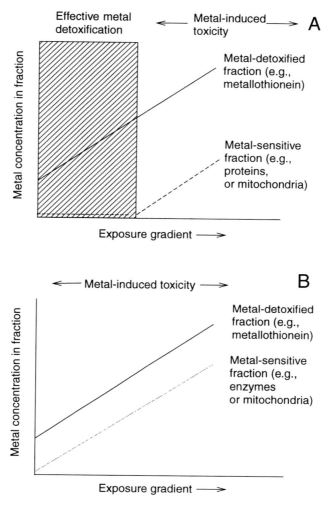

Figure 2 *Hypothetical relationships between chronic metal exposure and subcellular metal partitioning: (A) initial protection of metal-sensitive compartments up to a threshold metal exposure; (B) linear accumulation in the metal-sensitive compartments, without any threshold. Adapted from Campbell et al.[24] with permission from Aquatic Toxicology, Elsevier*

organisms (a freshwater bivalve, *Pyganodon grandis*, and yellow perch, *Perca flavescens*), collected from lakes representing a marked cadmium exposure gradient, the model predictions were not borne out; there was no threshold exposure concentration below which Cd spillover into potentially metal-sensitive fractions did not occur.[24] As internal Cd levels increased in response to the Cd exposure gradient, Cd concentrations tended to increase progressively in many sub-cellular compartments, even in specimens collected from sites located towards the low end of the metal exposure gradient. For the putative metal-sensitive compartments (organelles and heat-sensitive proteins), there was no clear evidence for a "hockey-stick" response as animal Cd concentrations increased. Analogous results were recently reported for benthic invertebrates collected along a metal concentration gradient in the Clark Fork River, Montana, USA. Cadmium concentrations increased progressively along the metal exposure gradient, in both the "metal-detoxified" and the "metal-sensitive" fractions, without any suggestion of a lower threshold below which the metal-sensitive fractions were protected.[29] It remains to be seen if this apparent absence of an exposure threshold can be transposed to chronically exposed organisms in general.[30]

The concept of metal speciation within organisms can be taken further, to include the trophic transfer of metals to predators. For example, recent experiments on the trophic transfer of Cd and Zn from two bivalves to their predators suggested that the metals present in the MSF fraction and bound to MT represent trophically available metal (TAM), whereas metals present in granules do not.[31] Increased partitioning of Cd to the TAM compartment within prey exposed to environmentally realistic external Cd concentrations enhanced Cd exposure along food chains.[32]

4.1.4 Essentiality

A number of metals are essential for biological functions and are critical to enzymatic and metabolic reactions within organisms. Essential elements are consistently present in healthy organisms and, if not present in sufficient quantities, result in deficiency symptoms that can be rectified by addition of sufficient quantities of the essential element. There is an optimal concentration range for each essential element to allow for normal metabolic functioning, and organisms will tend to regulate their internal concentrations of essential elements within this homeostatic range. Below this optimal range, toxicity occurs due to deficiency and has been observed in routine laboratory cultures[33]; above this optimal range, toxicity occurs due to excess. Some organisms will also regulate non-essential metals within a homeostatic range roughly equivalent to the natural background concentrations to which they have been exposed. When this range is exceeded, toxicity will occur.[2]

4.1.5 Acclimation, Adaptation, Tolerance

Because metals occur naturally in the environment, background concentrations are not negligible. In some areas that are naturally rich in minerals (*e.g.*, areas subject to mining), background concentrations can be high enough to result in toxicity to previously unexposed organisms. However, within these enriched "metalloregions",[34] there may be selection for metal-resistant populations. Clearly, it is mechanistically easier for organisms to acclimate or adapt to contaminants that occur naturally and that have been a feature of their evolutionary history, rather than to man-made contaminants with which they have no evolutionary history.

Organisms vary in their tolerance to metals depending on their and their ancestors' pre-exposure to those metals. Tolerance can occur via two processes: acclimation (shifting of tolerance within genetically defined limits) or adaptation (modifications of tolerance by changes in heritable genetic material (mutations) or changes in the frequency of heritable genetic material (which existed prior to the elevation of metals) in the population). The former physiological process occurs within an organism during a portion of its lifetime; the latter genetic process is passed on from parents

Risk Assessment of Metals in the Environment

to offspring[35–37] and can include the ability to detect and avoid metal contamination.[38] Acclimation is an energetic process that can reduce overall fitness, whereas adaptation may or may not require energy and affect overall fitness.[39,40] In some cases accumulation and detoxification mechanisms can be sufficient to deal with metal influxes without selection for a metal-tolerant population.[41] It is likely that most cases of metal tolerance, whether genetic or not, also include detoxification mechanisms such as metal sequestration. For instance, Boisson and co-workers[42] demonstrated reduced bioavailability and toxicity of Cd in oysters related to pre-exposure to chronic Cd contamination, and suggested that " ... regulatory thresholds ... should be reconsidered and should take into account the level of Cd already detoxified by the oysters through complexation processes."

Similarly, behavioural or other modifications can also help organisms deal with metals contamination without the need for adaptation. Some aquatic oligochaetes transfer excess metal to their anterior segments and then remove those segments and their accompanying metal loads via autotomy.[43] The response of plants to copper in soils can vary between adults and cuttings; the adults appear to extend their roots into deeper soil layers, avoiding contaminated soils and reducing exposure, which is not possible for the cuttings.[44] In estuarine environments, euryhaline crustaceans that show plasticity in Ca homeostasis are more likely to survive Cd exposure than are stenohaline species.[45] In freshwater, yellow perch (*P. flavescens*) are more tolerant to copper than are rainbow trout because they lose less sodium upon exposure to copper.[46]

Metal tolerance may or may not result in reduced genetic diversity[47] or in energetic costs. The latter are more likely for acclimation than for adaptation, but the energy balance of an organism is not a "zero-sum" exercise, *i.e.*, the pool of energy available is not necessarily finite. Up-regulation of a gene (*e.g.*, as a regulatory mechanism in response to increased or decreased metals exposure outside the normal range of experience) will not necessarily divert energy from a finite resource that cannot be expanded. There are many reasons why an organism would acclimate or adapt, including strategies to provide access to new energy reserves, to reduce competition by other species for more-available resources in metal-rich areas, to avoid predation, and to increase chances for the survival of offspring. Thus, acclimation and adaptation are normal responses of the organism to adjust the boundaries of its ecological niche to maximize its chances to survive and reproduce. The costs for tolerance will be variable, as will the potential consequences of subsequent exposure to other stressors. From an ERA perspective the key issue is whether or not metal-tolerant communities provide the same or similar ecological goods and services as metal-intolerant communities.

4.1.6 Mixtures

Contaminated environments typically comprise a mixture of organic and inorganic contaminants. Presently it is not possible to predict interactions among metals, let alone interactions between metals and other contaminants. However, consideration must be given to co-occurrence of contaminants as an explanation for observed toxicity where correlative rather than cause-and-effect studies are conducted. Less than additive and more than additive responses for metal–metal interactions are about as likely as strictly additive responses.[48] Simulation of generic metal interactions at epithelial-binding sites suggests that more than additive responses will be observed at low metal concentrations, additive responses at intermediate metal concentrations and less than additive responses at high metal concentrations, confirming that assumptions of additivity do not always apply.[49]

4.2 Aquatic Systems

In the preceding discussion we considered the general problems that arise when conventional ERA methods are used to assess the risk posed by metals in the environment. In the present section we focus on the aquatic environment and consider specific challenges associated with the risk assessment for metals in aquatic systems.

4.2.1 Exposure Assessment

Many of the metals that are of current environmental concern can be considered "data-rich" contaminants, for which environmental concentrations are reasonably well established. Note however that this is a fairly recent development; many of the historical data for trace metal concentrations in water proved unreliable, due to inadvertent contamination of the samples during or after collection, and due to filtration artefacts.[50,51] With the introduction of "trace-metal-clean" techniques, estimates of true dissolved metal concentrations have dropped very significantly.[52–55] However, geographical (spatial) data remain patchy and there is a need for better sampling strategies, analogous to those developed by the U.S. EPA in the late 1980s to derive regional estimates of lake chemistry and quantify the number of lakes that were vulnerable to acid precipitation.[56]

In addition to reliable estimates of current ambient metal concentrations, ERA for metals also requires environmental fate models that can be used to predict metal concentrations in various environmental compartments as a function of metal loading from natural and anthropogenic sources. Such models are available for organic contaminants, and indeed are widely used for ERA, but the prediction of environmental metal concentrations is still problematic.[3,57] It could be argued that this shortcoming is not a major problem for risk assessment *per se*, at least for the data-rich metals (since their ambient concentrations are reasonably well known), but it does constitute a major obstacle for the risk assessment of data-poor metals. In addition, such models are needed for risk *management*, e.g., for deciding whether reducing a particular metal input will in fact result in a meaningful decrease in the ambient metal concentration.

4.2.2 Effects Assessment – Water

Major progress has been achieved in recent years towards the goal of being able to predict the effects of metals on aquatic biota. Central to this progress has been an improved appreciation of how metals interact with aquatic organisms. In the vast majority of cases, this interaction involves a ligand-exchange reaction at an epithelial surface, leading to the formation of a metal complex at the surface.

$$M(L) + X\text{-membrane} \leftrightarrow M\text{-}X\text{-membrane} + L \text{ (charges not shown for simplicity)}$$

The metal-accepting ligand on the biological surface (X-membrane) may be a metal-sensitive site, or a *trans*-membrane metal transporter that moves the metal into the intracellular environment. Normally this surface complexation reaction is more rapid than the subsequent metal internalization step or expression of metal toxicity. In such cases the biological surface reaches equilibrium with the exposure medium, and the concentration of the "M-X-membrane" species proves to be a good predictor of the biological effect. In turn, the concentration of this surface complex will vary as a function of the free-metal ion concentration in the exposure medium, $[M^{z+}]$.

This description of the interaction of a metal with a biological surface is obviously oversimplified, but it does serve to demonstrate the link between the speciation of a metal and its bioavailability, and the importance of the free-metal ion concentration (or activity) as a predictor of metal bioavailability. Within this general construct, metal complexation in the exposure medium should lead to a decrease in metal bioavailability, and indeed such decreases are normally observed.[58] Similarly, other cations (*e.g.*, H^+, Ca^{2+}, Mg^{2+}) would be expected to compete with M^{z+} for binding at the epithelial surface, affording a certain protection against metal toxicity; such protection is well recognized in the aquatic toxicology literature.[59] In recent years, these various effects have been integrated into what is commonly known as the biotic ligand model (BLM) (Figure 3).[60,61] The BLM is gaining widespread acceptance among the scientific and regulatory communities – it has received provisional approval by the U.S. EPA, and most recently approval in principle by the EU. In addition, Environment Canada has recently decided to incorporate the BLM approach into the development of Canada's new water quality guidelines for metals.

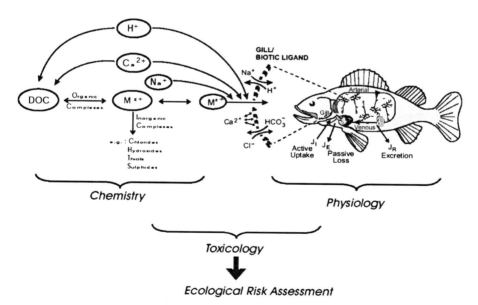

Figure 3 *Schematic representation of the BLM showing the interrelationships of water chemistry, respiratory physiology and toxicology. Adapted from Paquin et al.[60] With permission from Comparative Biochemistry and Physiology C, Elsevier*

To date, the BLM has been used for predicting the *acute* toxicity of metals in daphnid crustaceans and a few selected fish (rainbow trout, fathead minnow). Fortunately, cases of acute metal toxicity in aquatic systems are increasingly rare, and the focus of ERA has shifted more to chronic metal exposures. Extension of the present BLM approach to assess the chronic impacts of metals is, however, complicated by the fact that mechanisms of acute and chronic metal toxicity are not necessarily the same, as well as the fact that acute and chronic endpoints often differ.[62]

One of the more vexing problems to be overcome in moving from acute (short-term) to chronic (long-term) exposures is related to the acclimation phenomenon described earlier. The BLM implicitly assumes that the properties of the epithelial-binding site or "biotic ligand" are constant, unaffected by general water chemistry (pH, hardness) or by pre-exposure to metals. In fact, for fish and daphnids there is now convincing evidence that the metal-binding properties of the biotic ligand (*e.g.*, the apparent surface density of binding sites, their affinity for the metal of interest) *are* affected by these variables, and that the resulting changes significantly alter the metal sensitivity of the target organism.[63] These findings imply that a range of metal-binding properties for the biotic ligand will have to be incorporated into models for chronic toxicity.[61]

Another implicit assumption in the BLM approach is that metal toxicity can be predicted on the basis of waterborne exposures and that metals in food can be ignored. Several contemporary studies suggest that metal bioaccumulation[64] and toxicity[65] in aquatic animals can be strongly influenced by diet-borne metals.[66] A key ERA question is thus whether predictions derived solely on the basis of exposure to waterborne metals are sufficiently protective.[67] To answer this question one would need to be able to model the relative contributions of food and water to metal bioaccumulation for different exposure scenarios, and to be able to differentiate between food and water as sources of metal toxicity. This approach would advance the BLM to incorporate not only water chemistry and gill physiology, but also feeding, gut physiology and nutrition, as they apply to metal toxicology.

4.2.3 Effects Assessment – Sediment

Metals released into the aquatic environment tend to accumulate in sediments, and thus the prediction of the ecological effects of sediment-associated metals assumes great importance in ERA for metals. For organic contaminants in sediments, it has been conclusively demonstrated that pore water concentrations are the best predictors of bioavailability[68] and a similar approach has evolved for metals. It is assumed that pore water metal concentrations will reflect the metal's chemical potential at the sediment-water boundary, and that changes in this chemical potential will be related to the metal's bioavailability. Two approaches can be used to estimate metal concentrations in sediment interstitial water, $[M]_i$: one applies to oxic conditions and assigns the control of $[M]_i$ to sorption reactions on such sorbents as Fe- or Mn-oxyhydroxides or sedimentary organic matter[69]; the second applies to anoxic conditions and assumes that $[M]_i$ is controlled by precipitation-dissolution reactions with reactive amorphous sulfides (acid volatile sulfides (AVS)). This latter approach has been widely adopted for ERA purposes, particularly in the USA,[70] though not without controversy, as described below.

Sulfide is ubiquitous in aquatic sediments, as a result of the microbial sulfate reduction that occurs in oxygen-poor sediments, and the subsequent precipitation of sulfide with iron and other metals. By convention, cold extraction of sediments with 1 M hydrochloric acid yields a reactive fraction known as acid volatile sulfides, which includes amorphous FeS, mackinawite (FeS) and greigite (Fe_3S_4).[71] The basic concept is that many metals of environmental concern (Ag, Cd, Cu, Ni, Pb, Zn) have lower solubility products (i.e., are less soluble) than the amorphous iron sulfides present in sediments, and thus will tend to displace Fe, forming insoluble sulfides. Toxicity from these metals is not expected if there is sufficient sulfide available to bind all available metal; "available sulfide" is estimated by the AVS extraction, which also yields an estimate of the concentrations of those metals that were bound to AVS. This fraction, called simultaneously extracted metals (SEM),[†] will include not only metals that were bound to the extracted sulfide phases, but also any metals that were desorbed under the conditions of the cold HCl extraction. Provided the measured AVS concentration exceeds the concentration of SEM (SEM/AVS molar ratio <1), metal concentrations in sediment pore water should be very low and metal toxicity should not occur (i.e., toxicity to benthic organisms is assumed to be caused by dissolved metals only, with no contribution from the ingestion of particulate metals). On the other hand, metal toxicity is possible if the measured AVS concentration is less than the concentration of SEM; in such cases it becomes important to consider the magnitude of the concentration difference, i.e., not the SEM/AVS ratio but rather the difference [SEM] − [AVS]. Because of competition among the SEM for AVS, the metals of immediate concern will be those with the higher solubility products ($K_{Ag2S} < K_{CuS} < K_{PbS} < K_{CdS} < K_{ZnS} < K_{NiS}$) and those present at higher concentration. Note also that phases other than AVS, such as amorphous iron oxyhydroxides or particulate organic matter, may play a role in controlling pore water metal concentrations[72]; in such cases, the concentrations of metals that are in "excess" with respect to AVS will be lower than those predicted from the simple SEM–AVS model.

The SEM–AVS model has proven to be controversial and has provoked considerable debate. For example, geochemists note that AVS concentrations vary temporally and spatially, and question how these variations affect the SEM–AVS relationship.[73] Benthic ecologists, for their part, have demonstrated that many sediment-dwelling invertebrates create their own oxygenated burrows and derive their metal body burdens not from the sediment at all but from the overlying water compartment.[74] There are also legitimate questions about the contribution of ingested sediments to toxicity.[67,75] Despite these shortcomings, however, tests of the model on field-collected sediments and on metal-spiked sediments have consistently shown excellent agreement (>90%) between predictions and observations of the *absence* of toxicity (Figure 4).[76–79] Predictions of the occurrence of toxicity are admittedly much less accurate, presumably reflecting the existence of metal-binding phases other than AVS in many sediments. Indeed, the SEM–AVS model has recently been adapted to include metal binding to particulate organic matter, and this modified version has proven to be significantly better at predicting metal toxicity in sediments where [SEM] > [AVS].[80]

[†] $\sum_i [SEM_i] = [SEM_{Cd}] + [SEM_{Cu}] + [SEM_{Pb}] + [SEM_{Ni}] + [SEM_{Zn}] + \frac{1}{2}[SEM_{Ag}]$

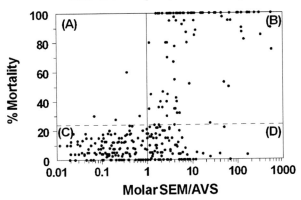

Figure 4 *Relationship between acute toxicity (defined as >24% mortality and shown by the horizontal dashed line) and the ratio of simultaneously extracted metals (SEM) to acid-volatile sulfide (AVS). A SEM/AVS value of 1.0, shown by the vertical solid line, is commonly used as a threshold separating toxic from non-toxic sediment samples. The quadrants represent areas where toxicity is correctly predicted (B), toxicity is incorrectly predicted (D), non-toxicity is correctly predicted (C) and non-toxicity is incorrectly predicted (A). Adapted from Shine et al.[78] With permission from Environmental Toxicology and Chemistry, Alliance Communications Group*

4.2.4 Indirect Effects of Metals

Traditional ERA methodology assumes that metals affect target species directly, either via waterborne or diet-borne metal exposure. It is however also possible that metal effects on a consumer organism may be *indirect*, *i.e.*, they may be mediated via the food web. A classic example of this type of ecotoxicological effect was observed in the whole lake acidification experiments carried out in the Experimental Lakes Area, Ontario, Canada.[81] The top predator in these lakes, the lake trout *(Salvelinus namaycush)*, was more or less insensitive to acidification *per se*, but its major prey items were progressively eliminated in the lakes as the pH declined, and the lake trout exhibited severe physiological and reproductive effects that could be linked to malnutrition. Similar examples can be found in the eutrophication literature,[82,83] but until recently food-web-mediated effects have not been considered in ERA for metals.

An example of the indirect effects of metals can be seen in recent studies on indigenous yellow perch (*P. flavescens*) in metal-contaminated lakes in eastern Canada.[84] Yellow perch in these lakes begin life feeding on zooplankton in the water column. During their second year of growth they shift to feeding on benthic invertebrates. Later in life their diet also includes smaller fish. Perch in the most metal-contaminated lakes do not undergo this normal sequence of diet shifts. Instead, they continue to utilize smaller prey throughout their lives because their primary prey species have been eliminated from the lakes, presumably by direct metal toxicity. The reduced availability of larger prey in lakes with high metal exposure leads to a sharp reduction in energy allocation to growth in adult perch, and results in stunting and a high degree of zooplanktivory at all ages.[85,86] A bioenergetic bottleneck results, with the perch's growth efficiency being reduced by the need to expend additional energy to catch and eat small-sized prey. Effectively, the following sequence of events occurs: metal exposure → reduced food abundance of certain dietary components → increased energetic costs of feeding → reduced growth efficiency. In summary, the invertebrate species that are most vital to growth and diet development of perch are among the most sensitive to metal contamination, and are among the slowest to recover. Yellow perch themselves tolerate a wide range of environmental conditions and are among the most widely distributed fish across the northern hemisphere. Thus, the "stunted perch" scenario that results from metal impacts on large macro-invertebrates is likely a widespread and common occurrence.

Functional redundancy is a common feature of aquatic ecosystems, where more than one species can play the same energetic role, for instance, fixing incident light energy (photosynthesis) or breaking down detrital organic matter (respiration). It is sometimes assumed that, because of functional redundancy, protecting ecosystem function will protect ecosystem integrity. However, this example clearly shows that this assumption does not universally hold. By changing ecosystem structure (the number and type of prey species present), metals have been shown to affect a higher consumer indirectly. The extent and significance of metal-mediated indirect effects need to be determined, particularly relative to inter-relationships between ecosystem structure and function.

4.3 Terrestrial Systems

Much of what has been described in the previous section for aquatic systems applies to terrestrial systems as well, namely, the need to identify which of the metal forms making up total soil metal are available for uptake by terrestrial organisms; the development of models that incorporate bioavailability into the prediction of metal toxicity and accumulation, under conditions of acute or chronic exposure; the consideration of exposure to diet-borne and soil pore water metals; and indirect effects of metals.

4.3.1 Exposure Assessment

Exposure assessment in terrestrial systems includes the determination of both metal *speciation* (in soil pore waters) and metal *fractionation*[8] (the distribution of a metal among various solid forms, and the prediction of the amount of metal that would be in solution). The determination of metal speciation in soil solution is fraught with the same difficulties that have been described earlier for aquatic systems, notably the need to account for the ubiquity of dissolved organic matter in the soil pore water. In addition, soils and soil pore waters are more spatially heterogeneous than are aquatic systems,[87] so the identification of a representative soil sample as well as the extraction of soil pore water are problematic. This latter challenge has confronted soil scientists for decades, but consensus as to the most appropriate method for metals has not yet been achieved.[88] Soil solution collection in the field using lysimeters is limited by both the time required to reestablish the soil structure after installation (at least one year) and the changes in solution chemistry that may occur during lengthy collection periods.[89] Approaches for extracting soil solution in the laboratory that are non-gravimetric or use dilute salt solutions likely result in much higher concentrations of dissolved metals than would occur in the soils in the field. A soil column leaching technique using dry soil has recently been demonstrated to produce dissolved metal concentrations similar to those observed in zero-tension lysimeters.[90,91] The free-metal ion concentrations in these solutions, after collection, can be determined by the same techniques used for samples from the aqueous environment (*e.g.*, ion-specific electrodes, anodic stripping voltammetry, ion-exchange resins or Donnan dialysis). Given the appropriate water quality data (major cations, major anions, pH, dissolved organic matter), the speciation of dissolved metals in soil pore water can in principle be calculated with the WHAM or NICA-Donnan chemical equilibrium models, but as discussed earlier for aqueous systems, the applicability of these models to natural samples remains to be demonstrated.

The fractionation of a metal (*i.e.*, its distribution among various solid forms) can be used to estimate the amount of metal that would be in soil solution, and is also key to predicting the potential for resupply of metal to the soil solution after depletion by plant uptake. It is more complex than fractionation in aquatic sediments, as soils are subject to periodic wetting and drying cycles, introducing a temporal variability that is largely absent in aquatic sediments. Metal fractionation data are available through experimentation, *e.g.*, by (sequential) chemical extractions or isotope exchange.[92,93] Sequential extraction identifies the amount of soil-metals removed from soil solids with reagents of increasing strength. While these "exchangeable pools" are not translatable into actual chemical species, the determination of a metal's "extractability" may nevertheless yield a better estimate of its bioavailability than does the analysis of the total soil-metal concentration; the more readily exchangeable pools are considered to be more bioavailable than those requiring stronger extractants. Dilute salt extraction (*e.g.*, $CaCl_2$) is frequently reported as the bioavailable fraction, and is more closely correlated with metal uptake or toxicity than is total. However, the $CaCl_2$-extractable metals are an imperfect estimate of bioavailable metals, as this approach does not consider the effect of pH, or competing cations (Ca, Mg) on metal uptake by organisms. Thus, $CaCl_2$ extraction may predict bioavailability better than total metal, among similar soils, but will not predict bioavailability well across different soils.[94–96]

4.3.2 Effects Assessment

From the perspective of most soil-dwelling organisms, both plants and invertebrates, metal speciation in the soil solution is thought to be more important in controlling metal uptake or toxicity than is metal fractionation in the (ingested) solid phase.[97,98] However, relationships between the speciation of a metal and its bioavailability are not straightforward for terrestrial organisms. Studies with higher plants have identified numerous exceptions to the original hypothesis that metal bioavailability could be predicted from the free-metal ion concentration, most of which can be attributed to the biological system's failure to conform to the assumptions associated with this hypothesis.[99–101] Some of these exceptions are addressed by the biotic ligand model (BLM), namely the influence of ions, including protons, that compete with the metal ion of interest both for root binding and for binding to sites on dissolved, colloidal or solid ligands in the exposure medium. Weng and co-workers[102] developed a predictive model for soil Ni phytotoxicity that accommodates the net effect of pH: the mitigation of toxicity that results from H^+ competing with Ni^{2+} for binding on the root, and the enhancement or toxicity that results from H^+ displacing Ni^{2+} from the soil solid fraction. Similarly, Thalaki and collaborators[103] have developed a terrestrial

BLM for Cu that predicts inhibition of barley root elongation in a range of soils amended with a soluble Cu salt. The model uses the chemical equilibrium model WHAM to predict the free Cu^{2+} concentration in the soil solution, using soil pH, the dissolved organic carbon concentration and the total soil Cu concentration as input parameters. As in the case of the nickel example, the copper model also accommodates the multiple roles of the H^+-ion, affecting the speciation of dissolved metals present in the soil solution and competing with Cu^{2+} for root binding leading to toxicity.

However, to characterise plant uptake of metals and their subsequent toxicity as a chemical equilibrium, as most current predictions of bioavailability (including the BLM) do, is an oversimplification. Living organisms may be in quasi-equilibrium with their environment, but plants grow; toxicity can cause changes in uptake and accumulation processes, and the organism itself may exert homeostatic control over metal distribution after binding, with the distribution pattern varying with species, or cultivars within species, as well as with exposure concentration. Metal concentrations in soil solution could be considered to exist in a quasi-equilibrium: the selective removal of ions from soil solution, as well as bulk water uptake to satisfy transpiration needs, leads to dissociation of complexes to restore equilibrium among free and complexed metals. Thus, it is not surprising that in a study of conditional stability constants for Cu accumulation in durum wheat (*Triticum aestivum* cv. durum) roots in hydroponic culture, over a wide range of Cu^{2+} activities, two types of root-binding sites were identified: in a log–log plot, high-affinity binding was not associated with toxicity and tissue accumulation was non-linearly related to exposure concentration, suggesting some homeostatic control over tissue accumulation, whereas low affinity binding was associated with toxicity, and was linearly related to Cu exposure.[104] Metal-binding sites in roots include the fixed negative charges on the molecules that dominate the water free space and Donnan free space, namely hemicelluloses, pectins, and galacturonic acids.[105] Binding of non-essential elements to these sites is likely not a primary cause of phytotoxicity, but its occurrence optimizes symplastic binding. Symplastic metal binding is more specific than apoplastic, and includes short-term associations, such as plasma membrane transporters, enzyme cofactors and long-term sequestration for homeostasis or detoxification, involving molecules such as phytochelatins and storage in the vacuole. Separation of apoplastic from symplastic binding of root metals has not often been addressed. Accumulation of Cd in roots has been demonstrated to be very sparingly reversible by exchange with Ca, and the conclusion is that most Cd associated with roots is symplastic, much of which is sequestered in vacuoles.[106] Characterising root-metal binding does not address the complexity of its fate thereafter. For example, Cd is readily transported from roots to shoots,[107] whereas Cu is translocated less readily; this difference in behaviour is partially attributable to copper's greater affinity for organic ligands, as well as to its essentiality, and the existence of mechanisms for homeostatic control. Even within species, root binding is not a good predictor of shoot metal accumulation. Among isolines of durum wheat, the partitioning of Cd between roots and shoots varies by a factor of 2[108]; cultivars of durum wheat with contrasting Cd accumulations in grain have been demonstrated to differ in mid-life translocation of Cd from root to shoot, rather than in root accumulation of Cd from solution.[109]

Characterising the metal-binding properties of roots is complicated by many observations of enhanced metal uptake at low free-metal ion activities in the presence of labile organic metal complexes. In several studies, calculations showed that diffusion of the free-metal ion through the boundary layer around the roots could not supply the amount of metal accumulated in the organism,[110] and that metal-ligand complexes could overcome this diffusional limitation by supplying the free-metal ion at the uptake site, by dissociation.[111] For example, in experiments run in nutrient solution in the absence of chelators, plant accumulation of Cd was linearly related to the free-metal ion activity in solution, with a slope of the relationship close to 1, as expected. However, at any given Cd^{2+} activity, uptake of Cd was greater in the presence of ligands than in the ligand-free controls (Figure 5).[112] Thus, there is often a kinetic component to metal accumulation by roots, in soil solutions with low metal ion activities. These considerations are probably more relevant to questions of micronutrient deficiencies than to metal toxicity, the latter normally being associated with higher metal levels.

Predicting metal activity in the boundary layer adjacent to the roots is complicated by the fact that it is under the influence of plant activity: roots exude organic molecules

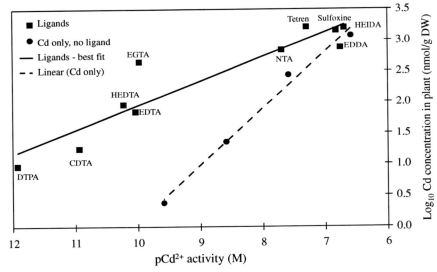

Figure 5 *Relationship between plant concentration (lettuce:* Lactuca sativa*) and the negative logarithm of solution Cd^{2+} activity in molar units (pCd^{2+}) in the absence of ligands (dotted line: Y = 9.22–0.92X; r^2 = 0.99, P<0.001) and in the presence of various chelators (solid line: Y = 5.9–0.4X; r^2 = 0.97, P<0.001). CDTA, trans-1,2-cyclohexyl-diamine-N,N,N',N'-tetraacetate; DTPA diethylenetriaminepentaacetate; EDDA, N,N-ethylenediaminediacetate; EDTA, ethylenediamine-N,N,N',N'-tetraacetate; EGTA, ethylene-bis-(oxyethylenenitrilo)-tetraacetate; HEDTA, N-2-hydroxyethyl-ethylenediamine-N,N,N',N',-tetraacetate; HEIDA, hydroxyethyliminodiacetate; NTA, nitrilotriacetate; sulfoxime, 8-hydroxyquinoline-5-sulphonate; Tetren, 1,4,7,10,13-pentaazatridecane. The concentration of all chelators is 50 µM, except Tetren (25 µM). Adapted from McLaughlin et al.[112] With permission from Plant Nutrition for Sustainable Food Production and Environment, Springer Science and Business Media*

such as siderophores (up to 20% of photosynthetically fixed carbon),[113] with attendant effects on the microflora in the root zone, and on the dissolved concentrations of essential metals such as Fe, Mn and Zn.[114] As well, plants may excrete H^+ depending on the source of N, and the resulting pH in the rhizosphere can be 1 to 1.5 units lower than in the surrounding bulk soil (with potentially a profound effect on metal speciation). Similarly, organic ligand content and composition in the soil adjacent to the plant roots will not be well represented by bulk soil measurements.[115] In a practical sense, measurements of rhizosphere soil are challenged by its small volumes, and the difficulty of its collection from field-grown plants.

As described earlier for aquatic organisms, direct uptake of lipophilic metal–ligand complexes by terrestrial organisms is also possible. In plants, intact metal–ligand complexes could also enter the apoplast directly without passing through a plasma membrane, through breaks in the endodermal barrier of the roots; such breaches have been observed in studies with very elevated concentrations of synthetic ligands, which destabilise membranes.[116] Direct uptake into the apoplasm as a result of disturbance has been studied using an apoplastic dye, 8-hydroxy-1,3,6-pyrenetrisulfonate (PTS), but a comparison of Cd and PTS uptake in young wheat plants, exposed to a solution in which most of the Cd was present as the free ion, suggested that apoplastic bypass accounted for less than 1% of the shoot accumulation of Cd.[117]

The speciation of metals in plants following uptake has the potential to influence both the trophic transfer of the metal and metal toxicity associated with ingestion of plant-based foods. Most research into trophic transfer of metals via the terrestrial environment has focused on those elements (such as Cd and As) that are not very phytotoxic, but are readily accumulated by edible plants.[118] Cadmium is readily accumulated by plants from soil, and particularly accumulates in grains, oilseeds and leafy green vegetables.[119] In plant tissues, Cd induces the production of phytochelatins; in grains, Cd is thought to be bound to phytates (myoinositol hexaphosphates), the principle storage form for P in seed.[105] *In vivo* Cd accumulation from lettuce by rabbits, or from grain by mice, in which the Cd had been incorporated during plant growth, has been compared with that for the same plant material to which Cd had been amended as a soluble salt; total accumulation in liver and kidney was less than 5% of total for all four diets, and slightly less accumulation was noted for Cd-incorporated plant material in liver (grain and mice) and kidney (lettuce and rabbits).[120,121] Other studies comparing diet materials likely to have different speciation of Cd in the tissues have suggested that speciation does influence accumulation in target organs,[122] although the presence of dietary fibres high in inositol hexa- and pentaphosphates, other nutritional elements and gastric pH are also important modifiers of Cd accumulation from food.[123,124]

5 Conclusions

As must be clear from the preceding sections, ERA for metals is not yet a mature field. After early delays, environmental regulators have accepted that ERA requires methods that are specific for metals; such methods are evolving rapidly, under the combined influence of both the regulators and the industries subject to regulation. In recognition of this fluid situation, we have divided this final section into two subsections: (i) recommendations applicable to present ERA and (ii) suggestions regarding the future evolution of ERA for metals.

5.1 Conclusions and Recommendations Applicable to Present ERA

- Traditional measures of persistence are clearly not useful for metals, since they offer no discriminatory power, *i.e.*, all metals are inherently persistent. Metals may, however, undergo various speciation changes, which in turn will have major effects on their geochemical mobility and their bioavailability.
- Such transformations among various metal species often involve changes in (weak) coordinate bonding, or changes in oxidation state; unlike the case for organic molecules, these changes are normally reversible, with the important consequence that the speciation of a metal is a function of the chemistry of the medium in which it is found. In contrast, the "speciation" of an organic contaminant is largely determined by the form in which the contaminant entered the environment and by any subsequent degradation processes.
- Since metal speciation can vary markedly from site to site, site-specific risk assessment will be more important for metals than for organic contaminants, for which a generic ERA may well suffice. Reliable estimates of ambient metal concentrations are required, as are estimates of the prevailing geochemistry (to allow informed speciation calculations).
- Given the pervasive influence of metal homeostasis and metal detoxification mechanisms, the bioaccumulation of a metal is rarely sufficient to predict environmental risk. It follows that neither BAFs nor BCFs are useful for ERA of metals; BCFs and BAFs are not an intrinsic property of the metal, but rather integrate the nature of the metal and the ambient exposure conditions.
- Environmental factors play an important role in determining the bioavailability of metals, in large part because the bioavailability of a given metal is so strongly influenced by its speciation, which in turn is largely determined by the chemical conditions in the receiving environment. For aquatic systems, the BLM has gained widespread acceptance among the scientific and regulatory communities; despite its limitations, the BLM clearly offers a much better estimate of metal bioavailability than does the traditional approach of measuring "recoverable metal" and then applying empirical corrections for the prevailing water hardness. However, the applicability of the BLM to terrestrial systems is still at an early stage and may be much more problematic.
- In addition to variations in metal bioavailability, ERA for metals must also consider the (variable) geological background concentrations of these naturally occurring substances, and their long-term influence on the indigenous fauna and flora. Behavioural and physiological tolerance must be taken into account; in the latter case, both acclimation and adaptation need to be considered, together with any energetic demands from such processes. A key issue is whether or not metal-tolerant communities provide the same or similar ecological goods and services as metal-intolerant communities.

- Metals typically occur in the environment as mixtures, not as individual metals – the conservative default of toxicity additivity may not always apply and should be assessed, not simply accepted.

5.2 Suggestions for Future Research

- The traditional "persistence" criterion works well in ERA for organic contaminants, but does not yield a meaningful hazard or risk ranking for metals. To develop a consistent approach for evaluating the persistence of organic and inorganic substances, persistence for both classes of contaminants could, in principle, be assessed with a common methodology in which mass balance models are used to determine the substance's behaviour in a specified environment or "unit world". Pioneering research in this area is currently being funded by the U.S. EPA.
- Given the marked influence of speciation on metal bioavailability, there is a clear need within ERA for validated chemical equilibrium models that can be used to predict metal speciation in the receiving environment. There is a critical need to ground-truth the existing models, *i.e.*, for measurements of the degree of metal complexation in minimally perturbed natural systems, and comparisons with the predictions of the WHAM and NICA-Donnan models.
- In addition to reliable estimates of current ambient metal concentrations, ERA for metals also requires environmental fate models that can be used to predict metal concentrations in various environmental compartments as a function of metal loading from natural and anthropogenic sources. Such models are available for organic contaminants but the prediction of environmental metal concentrations is still problematic. This shortcoming constitutes a major obstacle for the ERA of data-poor metals. In addition, such models are needed for risk *management*, *e.g.*, for deciding whether reducing a particular metal input will in fact result in a meaningful decrease in the ambient metal concentration and metal bioavailability.
- A variety of mechanisms exist for the detoxification of metals once they have entered a living cell, precluding simple predictions of metal-induced toxicity on the basis of metal quotas or burdens, as is done with organic contaminants. Determination of concentration–response relationships for bioreactive metals in tissues would be very helpful, especially for chronic exposures. The emphasis here must be on bioreactive metals in tissues, over and above concentrations necessary for survival and health in the case of essential metals.
- To date, the BLM has been used for predicting the acute toxicity of metals to various (aquatic) test species. With the focus of ERA shifting more to chronic metal exposures, there is a need for further development of the BLM to account for chronic toxicity (with the attendant challenges of acquired metal tolerance) and metal intake via food. This approach would advance the BLM to incorporate not only water chemistry and epithelial (gill) physiology, but also feeding, gut physiology and nutrition, as they apply to metal toxicology. In addition, the BLM should be extended to incorporate additional metals and organisms, and to consider saltwater exposures (not just freshwater).

- Given the success of the SEM–AVS model in predicting the absence of toxicity to benthic organisms, further research should be devoted to improve predictions of toxicity (rather that just its absence), and to determine why there exists an apparent disconnect between metal bioaccumulation and metal toxicity (*i.e.*, the current model predicts the absence of toxicity well, even in those cases where metal bioaccumulation occurs).
- Research on metals in the soil environment is presently lagging behind that in the aquatic environment. Reliable ERA methodologies are needed to deal with spatial heterogeneity in soils and the attendant scaling problems (landscape → field → plot → pot → rhizosphere). Other areas of concern include the impact of wetting and drying cycles in soils on metal bioavailability, and the contribution of dissolved metal complexes to metal accumulation and toxicity in the soil environment (where free-metal ion concentrations tend to be very low). Within the plant, there is a need to be able to distinguish between apoplastic and symplastic metal pools, to evaluate metal complexation in the symplasm, and to determine the role of each pool in metal toxicity, both to the plant and to consumers of the plants. In the ongoing development of a terrestrial BLM, the influence of plant physiology on metal accumulation and toxicity (and its modification by environmental factors) should be addressed.
- Finally, traditional ERA methodology assumes that metals affect target species directly, either via waterborne or diet-borne metal exposure. It is, however, also possible that metal effects on a consumer organism may be indirect, *i.e.*, they may be mediated via the food web. The extent and significance of such metal-mediated indirect effects in ecosystems, related to both structure and function, should be evaluated.

Acknowledgements

This work was carried out under the auspices of the Metals in the Environment Research Network (MITE-RN). PGCC is supported by the Canada Research Chair programme. We would like to thank Graeme Batley and Steve McGrath for helpful reviews of an earlier version of this paper.

References

1. US EPA, Framework for Ecological Risk Assessment, U.S. Environmental Protection Agency, Risk Assessment Forum Report No. EPA/630/R-92/001, Washington, DC, USA, 1992.
2. P.M. Chapman and F. Wang, *Human Ecol. Risk Assess.*, 2000, **6**, 965.
3. US EPA, Framework for Inorganic Metals Risk Assessment, U.S. Environmental Protection Agency, Risk Assessment Forum Report No. EPA/630/P-04/068B, Washington, DC, USA, 2004.
4. EURAS, Metal Risk Assessment Guidance Document: Background Document, EURAS, Ghent, Belgium, 2005.
5. EU, Technical Guidance Document in Support of the Commission Directive 93/67/EEC on Risk Assessment for New Notified Substances and the Commission Regulation (EC) 1488/94 on Risk Assessment for Existing Substances, Parts I-IV,

Office for Official Publications of the European Communities, Luxembourg, 1996.
6. S.P. Bradbury, T.C.J. Feijtel and C.J. Van Leeuwen, *Environ. Sci. Technol.*, 2004, **38**, 463A.
7. R.P. Schwarzenbach, P.M. Gschwend and D.M. Imboden, *Environmental Organic Chemistry*, Wiley, New York, NY 1993.
8. D.M. Templeton, F. Ariese, R. Cornelis, L.-G. Danielsson, H. Muntau, H.P. Van Leeuwen and R. Lobinski, *Pure Appl. Chem.*, 2000, **72**, 1453.
9. K. Simkiss and M.G. Taylor, in *Metal Speciation and Bioavailability in Aquatic Systems*, Vol. 1, A. Tessier and D. Turner (eds.), Wiley, Chichester, UK, 1995, 1.
10. A.Z. Mason and K.D. Jenkins, in *Metal Speciation and Bioavailability in Aquatic Systems*, Vol 10, A. Tessier and D. Turner (eds.), Wiley, Chichester, UK, 1995, 479.
11. D. Mackay, L.S. McCarty and M. MacLeod, *Environ. Toxicol. Chem.*, 2001, **20**, 1491.
12. D. Mackay, E. Webster, D. Woodfine, T.M. Cahill, P. Doyle, Y. Couillard and D. Gutzman, *Human Ecol. Risk Assess.*, 2003, **9**, 1445.
13. J.C. McGeer, K.V. Brix, J.M. Skeaff, D.K. DeForest, S.I. Brigham, W.J. Adams and A. Green, *Environ. Toxicol. Chem.*, 2003, **22**, 1017.
14. P.S. Rainbow, A. Geffard, A.Y. Jeantet, B.D. Smith, J.C. Amiard and C. Amiard-Triquet, *Mar. Ecol. Prog. Ser.*, 2004, **271**, 183.
15. P.M. Chapman, F.Y. Wang, C.R. Janssen, R.R. Goulet and C.N. Kamunde, *Human Ecol. Risk Assess.*, 2003, **9**, 641.
16. M. Haitzer, S. Hoss, W. Traunspurger and C. Steinberg, *Chemosphere*, 1998, **37**, 1335.
17. Y. Dudal and F. Gerard, *Earth-Sci. Rev.*, 2004, **66**, 199.
18. E. Tipping, S. Lofts and A.J. Lawlor, *Sci. Total Environ.*, 1998, **210**, 63.
19. C.J. Milne, D.G. Kinniburgh and E. Tipping, *Environ. Sci. Technol.*, 2001, **35**, 2049.
20. C.J. Milne, D.G. Kinniburgh, W.H. Van Riemsdijk and E. Tipping, *Environ. Sci. Technol.*, 2003, **37**, 958.
21. E. Tipping, C. Rey-Castro, S.E. Bryan and J. Hamilton-Taylor, *Geochim. Cosmochim. Acta*, 2002, **66**, 3211.
22. D.S. Smith, R.A. Bell and J.R. Kramer, *Comp. Biochem. Physiol. C - Toxicol. Pharmacol.*, 2002, **133**, 65.
23. W.G. Wallace, B.-G. Lee and S.N. Luoma, *Mar. Ecol. Progr. Ser.*, 2003, **249**, 183.
24. P.G.C. Campbell, A. Giguère, E. Bonneris and L. Hare, *Aquat. Toxicol.*, 2005, **72**, 83.
25. P.S. Rainbow, *Environ. Pollut.*, 2002, **120**, 497.
26. B.M. Sanders and K.D. Jenkins, *Biol. Bull.*, 1984, **167**, 704.
27. Y. Couillard, P.G.C. Campbell, J. Pellerin-Massicotte and J.-C. Auclair, *Can. J. Fish. Aquat. Sci.*, 1995, **52**, 703.
28. M. Baudrimont, S. Andres, J. Metivaud, Y. Lapaquellerie, F. Ribeyre, N. Maillet, C. Latouche and A. Boudou, *Environ. Toxicol. Chem.*, 1999, **18**, 2472.
29. D.J. Cain, S.N. Luoma and W.G. Wallace, *Environ. Toxicol. Chem.*, 2004, **23**, 1463.

30. R. Renner, *Environ. Sci. Technol.*, 2004, **38**, 90A.
31. W.G. Wallace and S.N. Luoma, *Mar. Ecol. Prog. Ser.*, 2003, **257**, 125.
32. D.R. Seebaugh, D. Goto and W.G. Wallace, *Mar. Environ. Res.*, 2005, **59**, 473.
33. B.T.A. Bossuyt and C.R. Janssen, *Environ. Toxicol. Chem.*, 2004, **23**, 2029.
34. M.J. McLaughlin and E. Smolders, *Environ. Toxicol. Chem.*, 2001, **20**, 2639.
35. C. Barata, D.J. Baird, S.E. Mitchell and A.M.V.M. Soares, *Environ. Toxicol. Chem.*, 2002, **21**, 1058.
36. D.E. Vidal and A.J. Horne, *Environ. Toxicol. Chem.*, 2003, **22**, 2130.
37. D.E. Vidal and A.J. Horne, *Arch. Environ. Contam. Toxicol.*, 2003, **45**, 184.
38. H. Lefcort, D.P. Abbott, D.A. Cleary, E. Howell, N.C. Keller and M.M. Smith, *Arch. Environ. Contam. Toxicol.*, 2004, **46**, 478.
39. J.B. Wilson, *Evolution*, 1988, **42**, 408.
40. L.T. Xie and P.L. Klerks, *Environ. Toxicol. Chem.*, 2004, **23**, 1499.
41. Q.X. Zhou, P.S. Rainbow and B.D. Smith, *J. Mar. Biol. Assoc. UK.*, 2003, **83**, 65.
42. F. Boisson, F. Goudard, J.P. Durand, C. Barbot, J. Pieri, J.C. Amiard and S.W. Fowler, *Mar. Ecol. Prog. Ser.*, 2003, **254**, 177.
43. D.E. Vidal and A.J. Horne, *Arch. Environ. Contam. Toxicol.*, 2003, **45**, 462.
44. M. Salemaa and S. Monni, *Environ. Pollut.*, 2003, **126**, 435.
45. J. Burke, R.D. Handy and S.D. Roast, *Environ. Toxicol. Chem.*, 2003, **22**, 2761.
46. L.N. Taylor, C.M. Wood and D.G. McDonald, *Environ. Toxicol. Chem.*, 2003, **22**, 2159.
47. J.D. Peles, W.I. Towler and S.I. Guttman, *Ecotoxicology*, 2003, **12**, 379.
48. W.P. Norwood, U. Borgmann, D.G. Dixon and A. Wallace, *Human Ecol. Risk Assess.*, 2003, **9**, 795.
49. R.C. Playle, *Aquat. Toxicol.*, 2004, **67**, 359.
50. H.L. Windom, J.T. Byrd, R.G. Smith and F. Huan, *Environ. Sci. Technol.*, 1991, **25**, 1137.
51. A.J. Horowitz, C. Lemieux, K.R. Lum, J.R. Garbarino, G.E.M. Hall and C.R. Demas, *Environ. Sci. Technol.*, 1996, **30**, 954.
52. K.H. Coale and R.H. Flegal, *Sci. Total Environ.*, 1989, **87/88**, 297.
53. J.O. Nriagu, G. Lawson, H.K.T. Wong and J.M. Azcue, *J. Great Lakes Res.*, 1993, **19**, 175.
54. G. Benoit, *Environ. Sci. Technol.*, 1994, **28**, 1987.
55. C.L. Creasey and A.R. Flegal, *Hydrogeol. J.*, 1999, **7**, 161.
56. D.H. Landers, W.S. Overton, R.A. Linthurst and D.F. Brakke, *Environ. Sci. Technol.*, 1988, **22**, 128.
57. P.R. Paquin, R.C. Santore, K.J. Farley, K.B. Wu, K. Mooney and D.M. Di Toro, *Review of Metal Fate, Bioaccumulation and Toxicity Models for Metals*, SETAC Press, Pensacola, FL, USA, 2005.
58. P.G.C. Campbell, in *Metal Speciation and Bioavailability in Aquatic Systems*, Vol. 2, A. Tessier and D. Turner (eds.), Wiley, Chichester, UK, 1995, 45.
59. J.S. Meyer, R.C. Santore, J.P. Bobbitt, L. Debrey, C.J. Boese, P.R. Paquin, H.E. Allen, H.L. Bergman and D.M. Di Toro, *Environ. Sci. Technol.*, 1999, **33**, 913.
60. P.R. Paquin, J.W. Gorsuch, S. Apte, G.E. Batley, K.C. Bowles, P.G.C. Campbell, C.G. Delos, D.M. Di Toro, R.L. Dwyer, F. Galvez, R.W. Gensemer, G.G. Goss, C. Hogstrand, C.R. Janssen, J.C. McGeer, R.B. Naddy, R.C. Playle, R.C. Santore,

U. Schneider, W.A. Stubblefield, C.M. Wood and K.B. Wu, *Comp. Biochem. Physiol.*, 2002, **133**, 3.
61. S. Niyogi and C.M. Wood, *Environ. Sci. Technol.*, 2004, **38**, 6177.
62. P.M. Chapman, *Mar. Pollut. Bull.*, 2002, **44**, 7.
63. S. Niyogi and C.M. Wood, *Human Ecol. Risk Assess.*, 2003, **9**, 813.
64. C. Munger and L. Hare, *Environ. Sci. Technol.*, 1997, **31**, 891.
65. S.E. Hook and N.S. Fisher, *Environ. Toxicol. Chem.*, 2001, **20**, 568.
66. N.S. Fisher and S.E. Hook, *Toxicology*, 2002, **181**, 531.
67. J.S. Meyer, W.J. Adams, K.V. Brix, S.N. Luoma, D.R. Mount, W.A. Stubblefield and C.M. Wood, *Toxicity of Dietborne Metals to Aquatic Biota*, SETAC Press, Pensacola, FL, 2005.
68. D.M. Di Toro, C.S. Zarba, D.J. Hansen, W.J. Berry, R.C. Swartz, C.E. Cowan, S.P. Pavlou, H.E. Allen, N.A. Thomas and P.R. Paquin, *Environ. Toxicol. Chem.*, 1991, **10**, 1541.
69. A. Tessier, D. Fortin, N. Belzile, R.R. De Vitre and G.G. Leppard, *Geochim. Cosmochim. Acta*, 1996, **60**, 387.
70. G.T. Ankley, D.M. Di Toro, D.J. Hansen and W.J. Berry, *Environ. Toxicol. Chem.*, 1996, **15**, 2056.
71. A. Tessier, R. Carignan and M.A. Huerta-Diaz, *Environ. Sci. Technol.*, 1993, **27**, 2367.
72. A. Tessier, in *Environmental Particles Environmental Analytical and Physical Chemistry Series*, J. Buffle and H.P. Van Leeuwen (eds.), Lewis Publishers, Boca Raton, FL, 1992, (11), 425.
73. J.W. Morse and D. Rickard, *Environ. Sci. Technol.*, 2004, **38**, 131A.
74. L. Hare, A. Tessier and L. Warren, *Environ. Toxicol. Chem.*, 2001, **20**, 880.
75. B.-G. Lee, S.B. Griscom, J.-S. Lee, J.J. Choi, C.-H. Koh, S.N. Luoma and N.S. Fisher, *Science*, 2000, **287**, 282.
76. D.J. Hansen, W.J. Berry, J.D. Mahony, W.S. Boothman, D.M. Ditoro, D.L. Robson, G.T. Ankley, D. Ma, Q. Yan and C.E. Pesch, *Environ. Toxicol. Chem.*, 1996, **15**, 2080.
77. W.J. Berry, D.J. Hansen, J.D. Mahony, D.L. Robson, D.M. Ditoro, B.P. Shipley, B. Rogers, J.M. Corbin and W.S. Boothman, *Environ. Toxicol. Chem.*, 1996, **15**, 2067.
78. J.P. Shine, C.J. Trapp and B.A. Coull, *Environ. Toxicol. Chem.*, 2003, **22**, 1642.
79. G.A. Burton, L.T.H. Nguyen, C. Janssen, R. Baudo, R. McWilliam, B. Bossuyt, M. Beltrami and A. Green, *Environ. Toxicol. Chem.*, 2005, **24**, 541.
80. D.M. Di Toro, J.A. McGrath, D.J. Hansen, W.J. Berry, P.R. Paquin, R. Matthew, K.B. Wu and R.C. Santore, *Environ. Toxicol. Chem.*, 2005, **24**, 2410.
81. D.W. Schindler, *Science*, 1988, **239**, 149.
82. R.S. Hayward and F.J. Margraf, *Trans. Am. Fish. Soc.*, 1987, **116**, 210.
83. J.S. Schaeffer, J.S. Diana and R.C. Haas, *J. Great Lakes Res.*, 2000, **26**, 340.
84. P.G.C. Campbell, A. Hontela, J.B. Rasmussen, A. Giguère, A. Gravel, L. Kraemer, J. Kovesces, A. Lacroix, H. Levesque and G.D. Sherwood, *Human Ecol. Risk Assess.*, 2003, **9**, 847.
85. G.D. Sherwood, J. Kovecses, A. Hontela and J.B. Rasmussen, *Can. J. Fish. Aquat. Sci.*, 2002, **59**, 1.

86. G.D. Sherwood, I. Pazzia, A. Moeser, A. Hontela and J.B. Rasmussen, *Can. J. Fish. Aquat. Sci.*, 2002, **59**, 229.
87. F.J. Caniego, R. Espejo, M.A. Martin and F. San Jose, *Ecol. Model.*, 2005, **182**, 291.
88. A. Lebourg, T. Sterckeman, H. Ciesielski and N. Proix, *J. Environ. Qual.*, 1998, **27**, 584.
89. W.H. Hendershot and F. Courchesne, *J. Soil Sci.*, 1991, **42**, 577.
90. J.D. MacDonald, N. Belanger and W.H. Hendershot, *Soil Sed. Contam.*, 2004, **13**, 361.
91. J.D. MacDonald, N. Belanger and W.H. Hendershot, *Soil Sed. Contam.*, 2004, **13**, 375.
92. Z.A.S. Ahnstrom and D.R. Parker, *Environ. Sci. Technol.*, 2001, **35**, 121.
93. A.S. Ayoub, B.A. McGaw, C.A. Shand and A.J. Midwood, *Plant Soil*, 2003, **252**, 291.
94. T.B. Kinraide, *Plant Physiol.*, 1994, **106**, 1583.
95. M.B. McBride, E.A. Nibarger, B.K. Richards and T. Steenhuis, *Soil Sci.*, 2003, **168**, 29.
96. T.B. Kinraide, *Plant Physiol.*, 1998, **118**, 513.
97. J.K. Saxe, C.A. Impellitteri, W.J.G.M. Peijnenburg and H.E. Allen, *Environ. Sci. Technol.*, 2001, **35**, 4522.
98. J.J. Scott-Fordsmand, D. Stevens and M. McLaughlin, *Environ. Sci. Technol.*, 2004, **38**, 3036.
99. E. Smolders and M.J. McLaughlin, *Soil Sci. Soc. Am. J.*, 1996, **60**, 1443.
100. E. Smolders and M.J. McLaughlin, *Plant Soil*, 1996, **179**, 57.
101. D.R. Parker, J.F. Pedler, Z.A.S. Ahnstrom and M. Resketo, *Environ. Toxicol. Chem.*, 2001, **20**, 899.
102. L.P. Weng, T.M. Lexmond, A. Wolthoorn, E.J.M. Temminghoff and W.H. Van Riemsdijk, *Environ. Toxicol. Chem.*, 2003, **22**, 2180.
103. S. Thalaki, H.E. Allen, D.M. Di Toro, A.A. Ponizovsky, C. Rooney, F.-J. Zhao and S. McGrath, *Proceedings of the 8th International Conference on the Biogechemistry of Trace Elements*, Adelaide, Australia, 2005.
104. P.M.C. Cypas Antunes, Beyond the free-ion activity model: The biotic ligand model for estimating copper uptake by durum wheat, Ph.D. Thesis, University of Guelph, Guelph, ON, Canada, 2004.
105. H. Marschner, *Mineral Nutrition of Higher Plants*, Academic Press, San Diego, CA, 1995.
106. W.E. Rauser, *Plant Physiol.*, 1987, **85**, 62.
107. I.S. Kim, K.H. Kang, K. Johnson-Green and E.J. Lee, *Environ. Pollut.*, 2005, **126**, 234.
108. N.S. Harris and G.J. Taylor, *BMC Plant Biol.*, 2004, **4**, 12.
109. D.Y. Chan and B.A. Hale, *J. Exp. Botany*, 2004, **55**, 2571.
110. E. Berkelaar and B.A. Hale, *Can. J. Botany*, 2003, **81**, 755.
111. H.P. Van Leeuwen, *Environ. Sci. Technol.*, 1999, **33**, 3743.
112. M.J. McLaughlin, E. Smolders, R. Merckx and A. Maes, *Plant Nutrition for Sustainable Food Production and Environment,* XIIIth *International Plant*

Nutrition Colloquium, T. Ando et al., (eds.), Kluwer Academic Publishers, Dordrecht, Netherlands, 1997, 113.
113. D.L. Jones, *Plant Soil*, 1998, **205**, 25.
114. J.A. Manthey, D.G. Luster and D.E. Crowley, Biochemistry of Metal Micronutrients in the Rhizosphere, CRC Press, Boca Raton, FL, 2004, 1.
115. V. Séguin, C. Gagnon and F. Courchesne, *Plant Soil*, 2004, **260**, 1.
116. A.D. Vassil, Y. Kapulnik, I. Raskin and D.E. Salt, *Plant Physiol.*, 1998, **117**, 447.
117. L. Van der Vliet, Cadmium accumulation in durum wheat: Roles of transpiration, dynamic steady state, life stage and apoplastic bypass, M.Sc. Thesis, University of Guelph, Guelph, Ontario, Canada, 2003.
118. M.J. McLaughlin, D.R. Parker and J.M. Clarke, *Field Crops Res.*, 1999, **60**, 143.
119. L. Friberg and M. Vahter, *Environ. Res.*, 1983, **30**, 95.
120. D.Y. Chan, W. Black and B. Hale, *Bull. Environ. Contam. Toxicol.*, 2000, **64**, 526.
121. D.Y. Chan, N. Fry, M. Waisberg, W.D. Black, and B.A. Hale, *J. Toxicol. Environ. Health A Curr. Iss.*, 2004, **67**, 397.
122. H.M. Crews, V. Ducros, J. Eagles, F.A. Mellon, P. Kastenmayer, J.B. Luten and B.A. McGaw, *Analyst*, 1994, **119**, 2491.
123. Y. Lind, J. Engman, L. Jorhem and A.W. Glynn, *Br. J. Nutr.*, 1998, **80**, 205.
124. M. Waisberg, W.D. Black, D.Y. Chan and B.A. Hale, *Food Chem Toxicol.*, 2005, **43**, 775.

Partitioning, Persistence and Long-Range Transport of Chemicals in the Environment

DONALD MACKAY, EVA WEBSTER AND TODD GOUIN

1 Introduction

Chemicals have played an important role in the development of our technologically dependent society, creating a wide range of economical and societal benefits. There has, however, been growing awareness that these same chemicals may pose risks to the environment and human health. Consequently, hazard assessment and risk management is needed to adequately control potential risks to humans and the environment.

Currently, the number of chemicals of commerce has been estimated to lie in the range 50,000 – 100,000. The European Chemicals Bureau, for example, lists 100,196 substances on its inventory of existing commercial substances.[1] Other substances such as the polycyclic aromatic hydrocarbons and chlorinated dibenzo-p-dioxins or "dioxins" are formed inadvertently as a result of human activities. Many chemicals such as pharmaceuticals, pesticides, and biocides are specifically designed to affect organisms in ways that are judged to be socially beneficial. Others are formed during industrial chemical synthesis as intermediates that are subsequently reacted to form a variety of chemical products. Inevitably, this large number of chemicals enters the environment as a result of deliberate or accidental synthesis, use, or accidental releases. It is recognised that some thousands of these substances are of sufficient concern that their fate in the environment should be carefully assessed and where necessary their discharges should be regulated, minimised, or eliminated. The United Nations Environment Program has identified 12 substances, commonly referred to as the "dirty dozen", which are deemed to be of sufficient concern that they should not be used, or at least that they should be severely restricted.[2] Others presumably deserve less regulation, thus there is a spectrum of concern dictated by the perceived risk that they pose to humans, wildlife, and ecological systems.

When assigning priorities to lists of chemicals of concern, the usual approach taken by regulatory agencies is to assess the chemical's hazard, *i.e.* its intrinsic potential to cause adverse effects regardless of the quantity used or emitted. Four such criteria are generally applied, persistence (P); bioaccumulation potential (B), which for organic substances proves to be largely a partitioning phenomenon between lipids and water or air; toxicity (T), which is dictated by biochemical considerations; and potential for long-range transport (LRT), which is of concern because it implies that chemicals used and released in one location may be transported to distant regions and there exert adverse effects. This has resulted in certain chemicals being classified as P, B, T, LRT substances or as persistent organic pollutants or POPs. Risk, however, results from a combination of hazard and quantity of chemical present resulting in exposure and thus the potential for adverse effects, *i.e.* risk is the product of hazard and exposure. This issue of quantity is recognised in the regulatory focus on "high production volume" (*i.e.* large production in terms of tonnes year^{-1}) or HPV chemicals. However, HPV does not necessarily translate into large emission rates, especially if the chemical is an industrial intermediate that is consumed during the production of other substances. The hazard criteria of P, B, T, LRT and perhaps HPV are only indicators of potential risk in that they each represent only parts of the overall picture. It is misleading to use them in isolation when managing chemicals for environmental purposes.

In this chapter, we focus on the issues of partitioning, persistence, and LRT since these characteristics of a chemical substance largely control how a given quantity or rate of chemical emission to the environment translates into exposures and hence effects.

Chemicals of environmental concern vary greatly in properties such as vapour pressure, solubility in water, and reactivity, thus their fate in the environment depends on the unique combination of persistence and partitioning, which differ considerably from chemical to chemical. For example, freons partition rapidly to the atmosphere and remain there for long periods of time due to their atmospheric stability, but polychlorinated biphenyls (PCBs) become sorbed to soils and sediments where they persist for several decades, trifluoroacetic acid partitions almost entirely into water, reacts only very slowly, and alkenes react rapidly and survive only for hours or days. These differences in properties translate into differences in chemical fate.

Examination of the chemical substances that have proved to be of primary concern reveals some common chemical characteristics including long environmental half-lives, especially attributable to slow reactions with hydroxy radicals in the environment; slow biodegradation rates in soils, sediments, and water; a sufficient vapour pressure that there is some partitioning to the atmosphere resulting in environmental transport; and hydrophobicity, namely, a strong tendency to partition into organic matter and lipids from water. Notable examples are halogenated alkanes and aromatics, phenols, polycyclic aromatics, pesticides containing halogen, phosphorus and nitrogen substituents, and some natural products that are toxic or persistent. The focus of this chapter is on organic substances since they are so numerous, but this does not imply a lower priority for metals, organo-metals, or other inorganic substances containing elements such as arsenic. These substances are generally addressed on a chemical-specific basis since the generalisations that apply to organic homologues do not apply to the same extent to inorganic or metallic substances.

2 Partitioning

2.1 Environmental Media

There are four primary environmental media between which chemicals partition, namely, the atmosphere (air), water (fresh and marine), soils, and sediments. Also important, at least for some substances, are aerosol particles, snow and ice, vegetation (both below and above ground), suspended particles or colloids in water, and a wide range of fauna including invertebrates, insects, fish, plankton, molluscs, birds, amphibians, mammals, and, of course that special mammal, humans. In most cases, environmental partitioning is dominated by the abiotic media, thus assessments of chemical fate often focus on how a mass of chemical partitions between these media. Partitioning to biota is usually addressed in a subsequent step. The volume of biotic media is usually small relative to abiotic media, but there are notable exceptions such as forests.

Table 1 lists a selection of these media and assigns to them order of magnitude volumes based on a surface land area of 100,000 km^2.

This area, which is about 300 km by 300 km, is sufficiently small that it can be geographically, climatically, and ecologically relatively homogeneous. It is about the area of Portugal and is about 40% of the area of the United Kingdom. Also included in this Table are representative volumes of certain parameters such as organic carbon and lipid contents, which prove to be valuable indicators of partitioning. These volumes and parameters are largely taken from the Equilibrium Criterion or EQC model of Mackay et al.[3]

2.2 Chemical Partition Coefficients

The objective is to measure or deduce partition coefficients or distribution coefficients between environmental media in order that relative concentrations and quantities of

Table 1 *EQC standard environment (Level III)*[3]

	Area (m^2)	Depth (m)	Volume (m^2)	Advective Residence Time (h)
Air	10^{11}	1000	10^{14}	100
Water	10^{10}	20	2×10^{11}	1000
Soil	9×10^{10}	0.2	1.8×10^{10}	
Sediment	10^{10}	0.05	5×10^8	50,000

	Volume fractions	Organic carbon or lipid mass fraction	Density (kg m^{-3})
Aerosol	2×10^{-11}		2000
Particles in water	5×10^{-6}	0.2	1500
Fish	10^{-6}	0.5	1000
Soil pore air	0.2		
Soil pore water	0.3		
Soil solids	0.5	0.02	2400
Sediment pore water	0.8		
Sediment solids	0.2	0.04	2400

chemical can be estimated. These partition coefficients are effectively applications of the Distribution Law of Nernst and Bertholet, who in the 1870s showed that the ratio of concentrations between immiscible phases, *i.e.* the partition coefficient, is relatively constant, especially under the dilute conditions which normally apply environmentally. Henry's Law applies the same concept to gas–liquid systems. It is essentially an assertion that at high dilution, the solute's activity coefficients in each phase are relatively constant.

Given the large number of chemicals and media, and differences in media composition and temperature, it is convenient to devise methods by which fundamental physico-chemical partition coefficients can be measured accurately and reproducibly in order to be used as surrogates for estimating the corresponding environmental quantities. It transpires that the phases illustrated in Figure 1 are useful for this purpose, namely, a gaseous phase (air), water, *n*-octanol (as a surrogate for organic matter), and the pure chemical phase, which may be liquid or solid (or occasionally vapour) at 25°C.

Notable are the air–water partition coefficient, K_{AW}, or Henry's Law constant, the octanol–water partition coefficient, K_{OW}, and the octanol–air partition coefficient, K_{OA}. If a pure chemical phase is present and saturation conditions exist, the corresponding partition coefficients can be regarded as the ratios of solubilities in the media. In water and octanol these are the actual solubilities. The solubility in air is

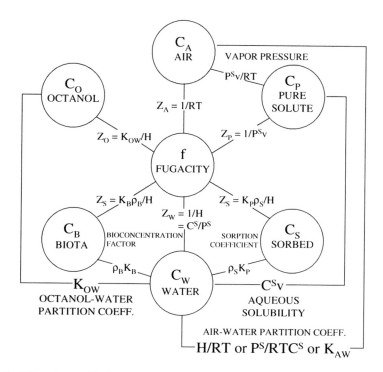

Figure 1 *The relationship between pure substance properties, partition coefficients, and Z values, from Mackay [9], reproduced with permission from Lewis Publishers*

essentially the vapour pressure, the concentration C (mol m^{-3}) being (from the ideal gas law) P/RT, where P is the vapour pressure (Pa), R the gas constant (Pa m^3 mol^{-1}), and T the absolute temperature (K). The solubility in octanol and water may or may not be measurable because of mutual miscibility.

These solubilities and partition coefficients are usually measured by a "shake flask" or equivalent "generator column" technique as reviewed by Mackay et al.[4], but increasingly reliance is being placed on GC or LC retention time methods in which a large number of determinations can be made rapidly by assuming a linear relationship between the partition coefficient and the retention time.[5-8] If solubilities or vapour pressures are estimated by this method, the values correspond to the liquid state chemical even if the substance is solid. For this and other reasons, it is also useful to deduce (for solids) the solubilities and vapour pressures of solids in their hypothetical sub-cooled liquid state. The ratio of solid-to-liquid values is the fugacity ratio, F, which can be estimated using Walden's rule (i.e. the entropy of fusion is approximately 56 J mol^{-1} K^{-1}) using the relationship

$$F = e^{6.79(1 - T_M/T)} \tag{1}$$

where T_M is the melting point temperature of the substance and T the temperature of the environment. It is these liquid state properties that can be correlated against molecular structure. The solid state values also depend on crystal structure and stability and melting point.[9]

In the case of dissociating or ionising organics such as carboxylic acids, phenols, and amines, it is essential to include consideration of the extent of ionisation as controlled by the acid dissociation constant (pKa) and the pH of the system.

In addition to these volumetric partition coefficients, there may be a need to define partition coefficients to interfaces such as the air–water interface in the form of ratios of interfacial concentrations (g m^{-2}) to bulk concentrations (g m^{-3}). These partitioning processes are particularly important when the area to volume ratio is large, i.e. the linear dimension of the phase is small as applies to fog droplets and snow flakes.

These coefficients are, of course, temperature dependent, however, a common or standard temperature of 25 °C is often assumed, this being convenient for laboratory determinations. Ideally, partition coefficients between these media should be measured over a range of temperatures, and correlations can be sought against molecular structure. Their temperature dependence is dictated by the enthalpy change experienced by the solute during phase-to-phase transfer. Generally, liquid-to-liquid transfer such as octanol–water involves a low energy change and temperature dependence is small, but for transfer to gaseous phases, the energy change can be substantial and the partition coefficients are correspondingly very temperature sensitive. It is very important to address this dependence when considering chemical fate, for example, in hot or cold climates, or at high altitudes.

2.3 Partition Coefficients in Fugacity Terms

When a chemical partitions and achieves equilibrium between two or more phases, the concentrations can be related using a partition coefficient or both concentrations can be converted to an equilibrium criterion which applies to both phases. The fundamental criterion is chemical potential, but it is more convenient to use activity or fugacity since at high dilution they are linearly related (approximately) to concentration. For environmental calculations fugacity proves to be particularly convenient. Concentration, C (mol m^{-3}), and fugacity, f (Pa), can be related through a proportionality constant or fugacity capacity, or Z value, Z (mol m^{-3} Pa^{-1}).

$$C = Zf \tag{2}$$

Z depends on the chemical, the medium, temperature, and only very slightly on pressure. If two concentrations, C_1 and C_2, are in equilibrium then

$$K_{12} = C_1/C_2 = Z_1 f/Z_2 f = Z_1/Z_2 \tag{3}$$

Essentially, the partition coefficient K_{12} is Z_1/Z_2 or each Z is "half" the partition coefficient, each expressing the non-ideality in only one phase. The Z values can be calculated from physical chemical principles or from empirical partition coefficients as shown in Figure 1. The equations for Z values are also given in Table 2, it being apparent that the partition coefficients are the corresponding ratios of Z values.

It is noteworthy that these partition coefficients are dimensionless volumetric values, *i.e.* ratios of g m^{-3} or mol m^{-3} in both phases. It is often preferable to express partition coefficients applicable to solid phases on a mass basis, *e.g.* (mg kg^{-1} soil)/(mg L^{-1} water), thus the partition coefficient has dimensions of L kg^{-1}. To convert to a dimensionless value requires that this value be multiplied by the solid phase density (kg L^{-1}).

Suggested relationships between the physico-chemical properties and Z values are listed in Table 2.

2.4 Quantitative Structure Property Relationships

Because free energies, activity coefficients, and hence physico-chemical properties are observed to vary systematically with changes in molecular structure, it is possible

Table 2 *Summary of Z value definitions*[9]

Compartment	Definition of Z mol m^{-3} Pa^{-1}	
Air	$1/RT$	$R = 8.314$ Pa m^3 mol^{-1} K^{-1}
		T = temperature, K
Water	$1/H$ or	H = Henry's Law constant, Pa m^3 mol^{-1}
	C^S/P^S	C^S = aqueous solubility, mol m^{-3}
		P^S = vapour pressure, Pa
Solid sorbent (*e.g.*, soil, sediment, particles)	$K_P \rho_S/H$	K_P = solid–water partition coefficient, L kg^{-1}
		ρ_S = density of solid, kg L^{-1}
Biota	$K_B \rho_B/H$	K_B = biota–water partition coefficient, L kg^{-1} or bioconcentration factor (BCF), L kg^{-1}
		ρ_B = density of biota, kg L^{-1} (often assumed to be 1.0 kg L^{-1})
Pure solute	$1/P^S v$	v = solute molar volume, m^3 mol^{-1}

to relate these properties to molecular structure and thus develop techniques for interpolation and modest extrapolation among homologues. This has been an active and fruitful area for both fundamental and environmental chemical research and a valuable correlative and predictive capability has developed. These quantitative structure property relationship (QSPR) methods (also known as quantitative structure activity relationships or QSARs) are particularly useful for checking that empirical data are in reasonable accord with other data for similar compounds and they are widely used to predict (with appropriate caution) properties of substances for which no data exist.

In their simplest form, they involve plots or regressions of partition coefficients or solubilities for members of a homologous series as a function of a molecular descriptor. This descriptor can range from a simple count of atoms such as number of chlorines in a series such as the chlorobenzenes to molar volume, molecular surface area, and various other quantities that can be calculated by molecular modelling or computational techniques. These QSPRs can be extended to treat several homologous series by introducing parameters for a variety of substituents such as halogens, methyl or methylene groups, alcohols, phenolic groups, ether linkages, and nitrogen-containing moieties. These "group contribution" methods and related "bond contribution" methods take the form of multi-parameter regressions and may involve corrections for proximity between groups and *cis–trans* and other isomerisation.[4] Another popular approach has been the use of molecular connectivity indices, which describe the topology of the molecule by counting atoms and their connections.[10] An example of a set of such QSPRs is the EPI Suite program, which is available free on the internet from the US EPA (http://www.epa.gov/opptintr/exposure/docs/episuite.htm) and the Syracuse Research Corporation (http://www.syrres.com). Many programs have been published in the literature and may be available from the authors. Others are available commercially. The reader can obtain information on these programs from the biennial QSAR conferences (*e.g.* Breton *et al.*[11]) or from the review edited by Walker[12] or the texts by Baum[13] and Mackay *et al.*[14]

Recently, in an attempt to improve the accuracy of these QSPRs, increasing attention is being devoted to "Poly-Parameter Linear Free Energy Relationships" (PPLFERs) in which, in principle, the free energy changes corresponding to the phase-to-phase transfer of a chemical are regressed against a number of parameters that are characteristic of the chemical. Abraham *et al.*[15] have pioneered this approach (which is widely used in medicinal chemistry) and others including Goss;[16] Goss and Schwarzenbach;[17] Breivik and Wania[18] have applied this concept to environmental systems by defining five chemical-specific parameters that influence molecular interactions in the partitioning phases as follows.

- E is the excess molar refraction representing non-specific interactions such as London and Debye forces
- S represents electrostatic interactions, *i.e.* dipole–dipole interactions
- A describes the overall hydrogen bond acidity, *i.e.* H donor or electron accepting effect
- B describes the overall hydrogen bond basicity, *i.e.* H acceptor or electron donor effect
- V is McGowan's characteristic molar volume.

Partitioning, Persistence and Long-Range Transport of Chemicals 139

If the free energy change is assumed to be a linear function of these five parameters, then the logarithm of the partition coefficient will also have the same linear dependence on these molecular descriptors with addition of a constant, *c*. The partition coefficient is then correlated by

$$\log K_{ij} = eE + sS + aA + bB + vV + c \tag{4}$$

relationships of this type that have been developed for the octanol–water, organic carbon–water, air–water, soil–air, and interfacial partition coefficients. The fitted "system constants" are specific to the two phases and express the relative solvation properties. In their critical review of this subject, Nguyen *et al.*[19] conclude that the PPLFER approach is "slowly gaining acceptance for use in the context of environmental chemistry and contaminant fate modelling" and they point out that descriptors are widely available for pharmaceutical drugs that are of increasing concern environmentally.

Although there are extensive compilations of physico-chemical properties such as solubilities and partition coefficients based on well-established experimental techniques, there remains a need to obtain more accurate data for chemicals of environmental concern for all relevant media and over a range of environmental temperatures. The greatest challenges are for polar chemicals including dissociating substances, surfactants that may accumulate at interfaces, and compounds of high molar mass for which values can be extremely low or high and thus difficult to measure accurately. Partitioning to vegetation is particularly problematic. A combination of careful experimental determinations over a range of temperatures and exploration of QSPR approaches seems the preferred approach for generating the thermodynamically consistent data necessary for characterising environmental partitioning.

2.5 *Level I Calculations*

Using the relationships in Table 2, it is possible to explore how chemicals partition between environmental media in terms of both mass and concentration. Such explorations are termed Level I calculations and are entirely hypothetical in nature, since chemicals in the environment are rarely at equilibrium, but they do reveal aspects of the likely partitioning behaviour of chemicals in the real environment. Losses by reaction and advection (flow in air or water) are ignored.

Table 3 gives physico-chemical properties of four chemicals. Table 4 gives the result of Level I calculations based on the following mass balance in which a defined quantity of chemical is assumed to be present partitioning at equilibrium between the three media of air, water, and soil with a volume ratio of 11000:22:1 as in Table 1 but with no sediment present.

The mass balance equations are as follows.

$$M = m_1 + m_2 + m_3 = V_1C_1 + V_2C_2 + V_3C_3$$
$$= V_1C_1 + V_2C_1K_{21} + V_3C_1K_{31}$$
$$= C_1(V_1 + V_2K_{21} + V_3K_{31})$$
$$\text{or } M = V_1Z_1f + V_2Z_2f + V_3Z_3f$$
$$= f(V_1Z_1 = V_2Z_2 + V_3Z_3) \tag{5}$$

Table 3 Physico-chemical properties of four chemicals[49]

	Naphthalene	Anthracene	Pyrene	Phenol
Molar mass (g mol^{-1})	128.2	178.2	202.3	94.1
Melting point (°C)	80	216	156	41
Solubility in water (g m^{-3})	31	0.045	0.132	88,360
Vapour pressure (Pa)	10.4	0.001	0.0006	47
log K_{OW}	3.37	4.54	5.18	1.46
Half-life in air (h)	17	55	170	17
Half-life in water (h)	170	550	1700	55
Half-life in soil (h)	1700	5500	17,000	170
Half-life in sediment (h)	5500	17,000	55,000	550

Table 4 Level I partitioning results as a percentage of chemical in each medium

	Naphthalene	Anthracene	Pyrene	Phenol
Air (1)	73.8	2.4	0.14	1.0
Water (2)	8.5	3.1	0.74	96.6
Soil (3)	17.7	94.5	99.12	2.4
Total	100	100	100	100

Knowing the volumes, partition coefficients, or Z values, the concentrations (mol m^{-3}), fugacities, and the amounts in each compartment can be calculated. Such calculations show the likely relative concentrations between the phases, the absolute values being irrelevant because they depend on the arbitrarily selected value of M, and are useful guides to monitoring programs. They also show how the mass of chemical partitions, *i.e.* where most of the chemical mass will likely be found. It is noteworthy that, as shown in Table 4, chemicals display very different partitioning tendencies.

2.6 Chemical space plots

An enlightening method of displaying and comparing the partitioning characteristics of a number of chemicals is to use a chemical space plot. Two such methods are in current use. First is a plot of log K_{AW} vs. log K_{OW} as shown in Figure 2. If the partition coefficients are internally consistent,[20] *i.e.* if the effect of the mutual solubility of water and octanol on the chemical's solvation in water and octanol is neglected, then the diagonals shown in Figure 2 correspond to lines of constant K_{OA}, the octanol–air partition coefficient or K_{OW}/K_{AW}. The second method is a plot of log K_{AW} vs. log K_{OA} in which the diagonals correspond to lines of constant K_{OW}.

By locating the chemical on such plots it becomes apparent which environmental compartment, or compartments, will likely be most important in assessing the chemical fate of the substance. A Level I calculation, as described above, is used to identify regions corresponding to compartments into which a substance will most likely partition. The four example chemicals are shown on the log K_{OW} vs. log K_{AW} plot in Figure 2. The environment is divided into compartments of air, water, and octanol,

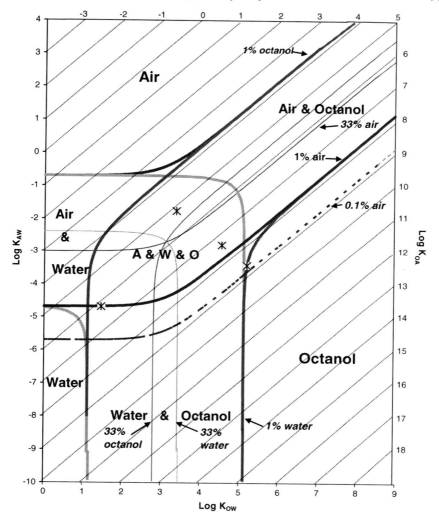

Figure 2 *Chemical space plot of log K_{AW} vs. log K_{OW} showing Level I partitioning between volumes of air, water, and octanol. The four sample chemicals are designated by an asterisk*

where the octanol compartment represents the organic carbon content associated with surface media, commonly represented by the soil and sediment compartments. For example, the partitioning between water and organic carbon, log K_{OC}, can be estimated as $0.35 \times \log K_{OW}$.[21]

Knowing the volume ratios of each environmental compartment, as well as the partition coefficients, K_{AW} and K_{OW}, it is possible to determine the mass fraction, F, of a chemical in each medium by

$$F_i = V_i K_{iW}/(K_{WW}V_W + K_{AW}V_A + K_{OW}V_O) \tag{6}$$

where the subscript i is air, water or octanol, and K_{WW} is 1.0. The volume ratios for air:water:octanol used by Gouin et al.[22], and adopted here, are respectively 650,000:1300:1.

It is important to note that the mass fraction of a substance for a particular environmental medium, as determined using Equation (6), will be strongly influenced by the volume fractions used. Thus, the partitioning behaviour of a substance will depend on the environment into which it is emitted, and may vary from region to region.

Summary

Partitioning of chemicals in the environment is controlled by the chemicals' intrinsic physico-chemical properties and by properties and volumes of the relevant environmental media. Partition coefficients or fugacity Z values can be used to describe this partitioning. Predictive methods are available and are becoming more reliable and applicable to a wider range of substances. The partitioning properties can be illustrated by simple Level I calculations and on chemical space plots. The primary needs are for more accurate data over a range of temperatures and for a wider variety of chemicals and environmental media.

3 Persistence

3.1 Chemical Reactivity

It transpires that the issue of evaluating persistence is largely that of determining rate constants or half-lives of degradation reactions in environmental compartments. These half-lives are, of course, specific to the chemical, and to the nature of the medium in which it is present and to the prevailing temperature.

These half-lives, $t_{1/2}$, can also be expressed as rate constants, k, where $t_{1/2}$ is ln $2/k$. Techniques for determining degradation rate constants are well established, especially for atmospheric reactions in which the primary mechanisms involve reactions with OH radicals, oxides of nitrogen, and ozone.[23] The reactions can be viewed as second-order with a first-order dependence on both the chemical concentration and the reactant concentrations. For example, it is common to express the rate of reaction with OH radicals using a second-order rate constant (e.g. cm^3 s^{-1} molecule^{-1}), and the prevailing concentration of OH radicals (e.g. molecules cm^{-3}), and the chemical concentration (e.g. g m^{-3}). The rate constant and the OH radical concentration can be combined to give a pseudo first-order rate constant (s^{-1}).

These rate constants can be related to molecular structure using techniques similar to that of QSPRs, and predictive methods have been developed, for example, the AOPWIN program based on Atkinson's data and estimation methods.[23]

More problematic are biodegradation rates, which are dependent on the prevailing microbial community as well as chemical structure with complications arising because of differences in degradation mechanisms under aerobic and anaerobic conditions, co-metabolism, and effects of temperature, nutrients, and toxicity. Again, predictive methods have been developed, usually based on estimates of the susceptibility of various groups to degradation.[24] There remains, however, difficulty in conducting reproducible laboratory scale tests that can be extrapolated to environmental conditions. A recent ECETOC review[25] discusses these issues in more detail as does the review by Jaworska *et al.*[26]

Many chemicals are subject to photolytic reactions either directly or indirectly following light absorption and activation by another substance. This topic has been reviewed by Mill[27] and in the text by Crosby[28] and estimation methods are available that take into account differences in local photolytic activity with latitude, time of year, and cloud cover.

Some chemicals are subject to hydrolysis under neutral, acid, or alkaline conditions, thus it is common to subject such substances to a systematic investigation of degradation over a range of pH to determine the corresponding rate constants with H_2O, H^+, and OH^- concentrations. A convenient review is that of Tratnyek and Macalady.[29]

Other mechanisms include oxidation by a variety of agents, reduction (especially under anaerobic conditions), and reaction with other substances that may be present in environmental media.

From the viewpoint of evaluating environmental persistence, the aim is to obtain estimates of half-lives for the subject chemical in all relevant media. These half-lives should include all the applicable reactions by adding the first-order rate constants to obtain an overall half-life. As with partitioning, it is important to include the temperature dependence in the form of an activation energy or energies. In many cases, the chemical of interest may degrade into other chemical species that are persistent, thus the total longevity of the reactant and products must be considered; the DDT-to-DDE conversion being an example.

3.2 Partitioning and Persistence

Persistence is most readily defined as the time required for a mass of chemical to be reduced to half its initial mass (a half-life, $t_{1/2}$) or the average time that a chemical mass remains in the environment (a residence time, τ). For a simple, first–order decay, the average residence time can be shown to be the reciprocal of the rate constant and thus $1/\ln(2)$ or 1.44 times the half-life. For a steady-state system with a constant chemical input rate, I (g h^{-1}), the residence time is the ratio of the mass in the system, M (g), to the input (and output) rate, *i.e.*, M/I (h). The evaluation of persistence is thus straightforward for a simple, first-order, one compartment system, but it can become more difficult for multi-compartment systems, as applies to the real environment.

The simplest regulatory approach for assessing persistence is to set criteria for degradation half-lives in each medium. For example, the Canadian approach is to declare a chemical as being persistent if any one of the following half-life criteria is exceeded.[30]

Half-life in air	2 days
Half-life in water	6 months
Half-life in soil	6 months
Half-life in sediment	1 year

Simple, single media assessments can be criticised in that a chemical may be unfairly penalised as being persistent by having a long half-life in a compartment into which it does not partition appreciably.

The influence of partitioning on the persistence of a chemical substance can be illustrated by the simple example depicted in Figure 3, in which the environment comprises two compartments, air and soil into which chemicals are discharged at a rate E (g h^{-1}). Thermodynamic equilibrium is established (or approached) by the chemical between these two compartments. The chemical is also subject to degrading reactions at rate constants specific to each medium, these rate constants being expressed for convenience as half-lives ($t_{1/2}$).

Assuming that losses are only by degrading reactions, a simple steady-state mass balance can be written, namely,

$$E = m_A k_A + m_S k_S \tag{7}$$

where m_A and m_S are the masses in air and soil, respectively and k_A and k_S (h^{-1}) are the corresponding reaction rate constants. The corresponding half-lives, $t_{1/2i}$, are $\ln 2/k_i$ or $0.693/k_i$. If the compartment volumes are V_A and V_S (m^3) then the concentration C_A is m_A/V_A and C_S is m_S/V_S (g m^{-3}). But at equilibrium, these concentrations are related by a dimensionless soil-air partition coefficient K_{SA} or C_S/C_A. The persistence, P, of the chemical, expressed as its residence time, is $(m_A + m_S)/E$ (h) or

$$P = (m_A + m_S)/(m_A k_A + m_S k_S) \tag{8}$$

or

$$1/P + F_A k_A + F_S k_S \tag{9}$$

where F_i is the fraction of the mass of the chemical in each compartment, for example, F_A is $m_A/(m_A + m_S)$.

Inspection of this equation reveals the following generalisations. The persistence of the chemical is an intrinsic or *intensive* property of the chemical and the environment, *i.e.* it is independent of the emission rate, E. It is controlled by the half-lives

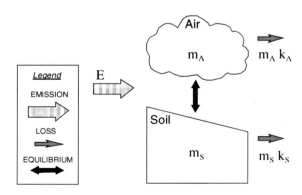

Figure 3 *Mass balance for a two-compartment (air–soil) Level II system*

(or rate constants) and how the chemical partitions (K_{SA} in combination with V_A and V_S). The persistence can be regarded as a weighted mean half-life, the weighting being on the basis of the partitioning characteristics of the substance.

As shown in Equation (9), the persistence of a chemical in a multimedia environment is thus a function of both its reactivity (expressed as a rate constant or half-life) and its partitioning tendencies.

Gouin et al.[22] have advocated a Level II approach similar to this for assessing persistence. In this case, a steady-state mass balance is set up for a four-compartment system in which there is an arbitrary constant chemical input or emission rate, equilibrium between compartments is assumed, and the concentration or the common fugacity adjusts to a value such that the total reaction rate equals the input rate. The applicable mathematics is described by Gouin et al.[22] and Mackay[9]. It can be shown that the overall half-life, $t_{1/2o}$ (which can be regarded as a metric of persistence) is given by

$$1/t_{1/2o} = \Sigma F_i t_{1/2i} \qquad (10)$$

where F_i is the fraction of the mass in each compartment as described above. Essentially, the rate constants are weighted in proportion to the respective masses. It then becomes clear which half-lives have the greatest significance and should be determined most accurately.

Focussing only on single media half-lives may also provoke regulatory demands for half-life data that are not really necessary. A response to this concern is the "realistic presence" concept that suggests that if a Level I or II calculation shows that less than perhaps 5% of the chemical mass partitions into a compartment, then an infinite half-life can be assumed with little effect on the resulting persistence, or the compartment can be ignored.[31]

An insightful simplification can be derived because partitioning of organic chemicals to soil and sediment is characterised by K_{OW}, the octanol–water partition coefficient. The environment can then be viewed, for partitioning purposes, as volumes of air, water, and octanol, the relative proportions described earlier, being 650,000, 1300, and 1, respectively. The masses of chemical in each, at equilibrium are then 650,000 K_{AW}, 1300 and 1.0 K_{OW} from which the fractions, F_i, can be calculated. If half-lives can be assigned to air, water, and the "octanol" phases, the overall persistence (in the form of a half-life) can be readily estimated using the above equation.

Calculations of overall half-life or persistence demonstrate the importance of both partitioning and reaction half-lives. The real environment is, of course, much more complicated containing more compartments that may or may not be at equilibrium or steady-state, but the broad principles governing partitioning and persistence still apply.

3.3 Level III as an Improved Estimate of Persistence

The major criticism of the Level II method of calculating persistence is that it assumes equilibrium to apply throughout the environment. This can be corrected by using a Level III calculation in which mass balance equations are set up for each well-mixed compartment and the equations solved for the corresponding concentrations or fugacities. Compartments are now generally not at equilibrium. Table 5 gives selected transport parameters in the form of velocities. The transport velocities comprise mass transfer coefficients characterising diffusive velocities (m h^{-1}) in air and in water at various phase boundaries, area-specific precipitation or deposition and resuspension rates, and area-specific water runoff rates (m^3 h^{-1} m^{-2}). These velocities vary considerably, both spatially and temporally, and those given in Table 5 are only representative values. Mackay[9] describes how these velocities are converted to transport rate expressions applicable to specific chemicals. It is again possible to calculate an overall reaction persistence or residence time as the mass in the system divided by the total reaction rate. If advection (flow) processes are included, an advective residence time can also be calculated. An overall persistence including losses by both reaction and advection can also be deduced.

It is noteworthy that when equilibrium conditions do not apply, as is generally the case, the overall reaction rate is no longer first-order and an overall half-life is irrelevant. Obviously, during a change in environmental concentrations resulting from changing rates of emission, some compartments will respond faster than others, so there is a spectrum of residence times depending on the distribution of the substance between compartments. The average or overall value of the residence time or persistence is still valid. Webster et al.[32] have discussed this issue in detail showing how Level III evaluations can lead to more meaningful estimates of persistence.

A typical Level III result is given in Figure 4 in which naphthalene is released into air and water at rates of 8 and 2 kg h^{-1}, respectively, for illustrative purposes. The resulting partitioning in Table 6 is clearly different from the equilibrium result and the results now depend on the "mode-of-entry" (MOE), i.e. the compartment into which the substance is emitted.

This MOE dependence is also illustrated in Table 6 where the chemical residence times are given for each emission separately and in combination. When emissions occur simultaneously, the result is the weighted sum of the emissions individually.

If desired, the time course of concentration changes can be deduced in detail by Level IV calculations in which a differential equation is set up for each compartment

Table 5 *Transport velocities (EQC standard environment)[3]*

	m/h	m/year
Air side air–water MTC	5	43,830
Water side air–water MTC	0.05	438.3
Rain rate	10^{-4}	0.8766
Aerosol deposition velocity	6×10^{-10}	5.26×10^{-6}
Soil air phase diffusion MTC	0.02	175.32
Soil water phase diffusion MTC	10^{-5}	0.08766
Soil air boundary layer MTC	5	43,830
Sediment–water MTC	10^{-4}	0.8766
Sediment deposition velocity	5×10^{-7}	0.004383
Sediment resuspension velocity	2×10^{-7}	0.001753
Soil water runoff rate	5×10^{-5}	0.4383
Soil solids runoff rate	10^{-8}	8.77×10^{-5}

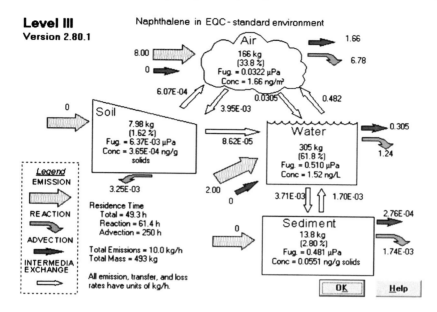

Figure 4 Level III model (version 2.80 available from http://www.trentu.ca/cemc) result for naphthalene emitted at rates of 8 and 2 kg h^{-1} to air and water simultaneously

Table 6 Partitioning and residence times or persistence of naphthalene in a non-equilibrium Level III system as shown in Figure 4 as a function of MOE

		MOE		
		8 kg h^{-1} to Air	2 kg h^{-1} to Water	Both
Mass (kg)	In air	157.1	9.3	166.4
	In water	4.3	300.3	304.6
	In soil	7.5	0.4	7.9
	In sediment	0.2	13.6	13.8
Mass (% of total)	In air	92.9	2.9	33.8
	In water	2.6	92.8	61.8
	In soil	4.4	0.1	1.6
	In sediment	0.1	4.2	2.8
Residence time (h)	Reaction	26.3	201	61.4
	Advection	107	822	250
	Overall	21.1	162	49.3

and the detailed dynamic response of the system to changing inputs is evaluated. Systems of such equations can be solved numerically[9] or analytically.[33] Fenner et al.[34] have shown that the steady-state residence time is equal to the "equivalence width", which is the area under the mass–time decay curve divided by the initial mass and is the average residence time. The same average persistence information can be gained from steady-state and dynamic simulations, but the dynamic simulation gives the additional distributions of persistences.

Stroebe et al.[33] have suggested that it is the slowest decay process that should be of most concern. They have introduced the concept of a "temporal remote state" describing the condition of the system long after emissions have stopped. This places most focus on the behaviour of highly persistent chemicals that have partitioned into media in which they react only slowly, usually sediments and soils.

Summary

In summary, there are several definitions of persistence and a variety of modelling techniques can be used to generate quantitative estimates. The use of models as distinct from empirical or monitoring investigations appears necessary because persistence is a quantity that cannot be measured directly in the environment. It is clear that persistence depends primarily on the reaction kinetics of the substance and on its partitioning characteristics, with the former being the more difficult to quantify. There is a compelling incentive to develop improved methods of measuring these kinetic processes and extrapolating rate data from laboratory data to environmental conditions.

4 Long-Range Transport

Partitioning and persistence processes play a key role in determining the potential of a substance to undergo LRT in air, ocean currents, and even in biota. This issue has ethical aspects because it is viewed as unacceptable that the beneficial use of chemicals in one community or nation results in contamination and adverse effects in another. Of particular concern is transport from populous industrial temperate regions to polar regions where the low temperatures tend to favour partitioning from the atmosphere to solid and liquid media and to reduce rate constants for degrading reactions, resulting in increased persistence and exposure. The "acid rain" and polar ecosystem contamination issues are of this type. International agreements to ban or regulate such substances[2,35] have been formulated in response to this issue.

Obviously, the substance must partition appreciably into a mobile medium such as air if it is to be transported in that medium. Further, it must persist long enough to survive the journey. Monitoring programs in remote ecosystems can identify chemicals that have been transported, but they cannot determine the relative quantities of different chemicals that have been transported and thus the inherent susceptibility of a chemical to undergo LRT. As in the case of persistence, it is possible for a substance to change its chemical identity during transport, thus consideration must also be given to the degradation products. As analytical methods improve to the picogram and femtogram levels, it seems likely that virtually all chemicals of commerce will be measurably present in all parts of the planet. Certainly, all persistent chemicals are already routinely detected in remote regions such as the Arctic and Antarctic.

There are two general approaches for assessing the potential for LRT. The first is to rely on large-scale regional and even global models of the types applied to acid rain. For instance, Global Circulation Models, describing atmospheric transport, have been developed and applied to contaminants using real or typical meteorological data.³⁶ Simpler multimedia models have also been developed and applied to substances such as the hexachlorocyclohexanes.³⁷⁻³⁹ These models can deduce source–receptor relationships, and are thus capable of assessing how chemicals differ in their potential for LRT. For example, Wania⁴⁰ has developed the concept of an Arctic Contamination Potential by running a model repeatedly for substances in a defined chemical space plot as described earlier. The partitioning characteristics that result in potential for LRT can then be determined.

The second approach is to conduct "evaluative" calculations in which the transport potential is compared in a common and simple model system. This has resulted in the concept of a "half distance" for transport⁴¹ and a "characteristic travel distance".⁴²,⁴³ These distances are deduced using an assumed velocity and taking into account losses during transport by degrading reactions and transport between media. For example, Beyer et al.⁴³ showed that the characteristic travel distance, CTD, in air can be estimated using a simple Level III model as

$$\text{CTD} = U T_R F_A \tag{11}$$

where U is the assumed air velocity, T_R the overall persistence of the chemical to reaction, and F_A the fraction of the mass of the chemical in air. In many respects this is analogous to the retention time in a chromatographic system, but in this case there can be degradation in both the mobile and stationary phases. Another notable approach has been the models developed by Scheringer and colleagues,⁴⁴⁻⁴⁶ who have advocated the concept of a "spatial range" as reflecting this transport potential.

A detailed account of these various methods is beyond the scope of this chapter, but it is sufficient to conclude that LRT potential can be quantified for real or evaluative systems provided that reliable data exist for the degradation rates and partitioning characteristics of the substance. Fenner et al.⁴⁷ have explored how various available models compare for estimates of both persistence and LRT and have shown that the results are highly correlated and are largely determined by chemical properties.

It is noteworthy that a number of factors, which may influence transport potential may be poorly characterised by the evaluative models that are currently available. Examples include chemicals of emerging concern such as the polybrominated diphenyl ethers, perfluorinated organics, and some current-use pesticides. Many models characterise precipitation as a steady-state, continuous event, however, it has been demonstrated that intermittent precipitation may significantly influence the mobility of some substances. For instance, Muir et al.⁴⁸ estimate that the CTD for a number of current-use pesticides, such as alachlor, atrazine, and metolachlor can be as much as 20 times greater when the models are adjusted to express precipitation as intermittent rather than continuous. These results suggest that when assessing the LRT potential for relatively reactive and water-soluble chemicals, the assessments should consider the LRT of these substances during periods of little or no precipitation.

Another issue relates to certain substances that are persistent and relatively non-volatile, for which particle-bound or aerosol transport may be the primary mechanism influencing fate and mobility. For these substances, the LRT will be similar to the transport potential of the particles with which they are associated. If meteorological conditions are such that the plume containing the aerosol is conveyed to high altitudes, there will be a reduced potential for deposition, and an increased potential for LRT by high velocity and high altitude air currents. It is thus likely that these particles can migrate to remote areas, such as the Arctic, in a relatively short period of time.

Summary

In summary, long-range atmospheric transport can be viewed and quantified by a combination of information on chemical persistence and partitioning and the movement of air masses, including aerosols. In principle, similar approaches can be applied to oceanic and freshwater transport and to transport in biota. With this information, the relative susceptibility of candidate substances to LRT can be evaluated and by exploiting atmospheric and oceanic circulation models estimates can be made of global-scale source–receptor relationships. There is a compelling incentive for a particular focus on transport to cold polar regions where chemicals are more persistent and indigenous populations are more dependent on local wildlife for sustenance.

5 Conclusions

The chemical principles that determine partitioning, persistence, and LRT potential play key roles in describing the fate, exposure, and effects of chemicals in our environment from local to national, continental, and global scales. Partitioning, persistence, and LRT can best be assessed by a combination of environmental monitoring, laboratory studies, and modelling techniques. The use of models as distinct from empirical or monitoring investigations appears necessary because both persistence and LRT are hazard metrics that cannot be measured directly in the environment. It is clear that persistence and LRT depend primarily on the reaction kinetics of the substance and on its partitioning characteristics, with the former being more difficult to quantify. Environmental parameters, such as temperature dependence, dry and wet particle deposition rates, rates of precipitation, and the transport of air and water masses also influence the fate and distribution of chemical substances. There thus remains a compelling incentive to develop improved methods of measuring these kinetic, thermodynamic partitioning processes, and extrapolating rate and equilibrium data from laboratory to environmental conditions.

Acknowledgements

The authors thank the Natural Sciences and Engineering Research Council of Canada (NSERC) and the consortium of chemical companies that support research at the Canadian Environmental Modelling Centre.

References

1. European Chemicals Bureau (ECB) website, http://ecb.jrc.it/, 2005.
2. UNEP, Final act of the conference of plenipotentiaries on the Stockholm Convention on persistent organic polutants, *UNEP/POPS/CONF/4*, United Nations Environment Programme, Stockholm, Sweden, 2001.

3. D. Mackay, A. Di Guardo, S. Paterson and C.E. Cowan, Evaluating the environmental fate of a variety of types of chemicals using the EQC model, *Environ. Toxicol. Chem.,* 1996, **15**(9), 1627–1637.
4. D. Mackay and R.S. Boethling, (eds), *Handbook of Property Estimation Methods for Chemicals: Environmental and Health Sciences*, CRC Press, Boca Raton, 2000, 1–481.
5. K.M. Ballschmiter, Analysis of polychlorinated biphenyls (PCB) by glass capillary gas chromatography. Composition of technical Aroclor and Clopen-PCB mixtures, *Zell. Fres. Z. Anal. Chem.*, 1980, **302**, 20–31.
6. X. Zhang, K.W. Schramm, B. Henkelmann, C. Klimm, A. Kaune, A. Kettrup and P. Lu, A method to estimate the octanol–air partition coefficient of semi-volatile organic compounds, *Anal. Chem.*, 1999, **71**, 3834–3838.
7. Y. Su, Y.D. Lei, G.L. Daly and F. Wania, Determination of octanol–air partition coefficients (K_{OA}) for chlorobenzenes and polychlorinated naphthalenes from gas chromatographic retention times, *J. Chem. Eng. Data*, 2002, **47**, 449–455.
8. D.A. Hinckley, T.F. Bidleman, W.T. Foreman and J.R. Tuschall, Determination of vapor pressures for nonpolar and semipolar organic compounds from gas chromatographic retention data, *J. Chem. Eng. Data*, 1990, **35**, 232–237.
9. D. Mackay, *Multimedia Environmental Models: The Fugacity Approach*, 2nd edn, Lewis Publishers, Boca Raton, 2001, 1–261.
10. L.B. Kier and L.H. Hall, *Molecular Connectivity in Structure–Activity Analysis*, Research Studies Press, Ltd., Letchworth, Hertfordchire, England, and Wiley, New York, 1986.
11. R. Breton, G. Schurmann and R. Purdy, *Proceedings of QSAR 2002*, QSAR and Combinatorial Science, 2003, **22**, 1–409.
12. J.D. Walker, (ed), Annual review: quantitative structure–activity relationships, *Environ. Toxicol. Chem.*, 2003, **22**, 1651–1935.
13. E.J. Baum, *Chemical Property Estimation: Theory and Application*, Lewis Publishers, Boca Raton, FL, 1997.
14. D. Mackay, E. Webster and A. Beyer, Defining the bioaccumulation, persistence, and transport attributes of priority chemicals, in *American Chemical Society Symposium Series No. 773 Persistent, Bioaccumulative, and Toxic Chemicals*, Vol II, R. Lipnick, B. Jansson, D. Mackay and M. Petreas (eds), ACS, Washington, 2001, 14–28.
15. H.M. Abraham, A. Ibrahim and A.M. Zissimos, Determination of sets of solute descriptors from chromatographic measurements, *J. Chromatogr. A*, 2004, **1037**, 29–47.
16. K.U. Goss, Free energy of transfer of a solute and its relation to the partition constant, *J. Phys. Chem. B*, 2003, **107**, 14025–14029.
17. K.U. Goss and R.P. Schwarzenbach, Linear free energy relationships used to evaluate equilibrium partitioning of organic compounds, *Environ. Sci. Technol.*, 2001, **35**, 1–9.
18. K. Breivik and F. Wania, Expanding the applicability of multimedia fate models to polar organic chemicals, *Environ. Sci. Technol.*, 2003, **37**, 4934–4943.
19. T.H. Nguyen, K.-U.Goss and W.P. Ball, Polyparameter linear free energy relationships for estimating the equilibrium partition of organic compounds

between water and the natural organic matter in soils and sediments, *Environ. Sci. Technol.*, 2005, **39** (4), 913–924.
20. A. Beyer, F. Wania, T. Gouin, D. Mackay and M. Matthies, Selecting internally consistent physical-chemical properties of organic compounds, *Environ. Toxicol. Chem.*, 2002, **21**, 941–953.
21. R. Seth, D. Mackay and J. Muncke, Estimating of organic carbon partition coefficient and its variability for hydrophobic chemicals, *Environ. Sci. Technol.*, 1999, **33**, 2390–2394.
22. T. Gouin, D. Mackay, E. Webster and F. Wania, Screening chemicals for persistence in the environment, *Environ. Sci. Technol.*, 2000, **34**, 881–884.
23. R. Atkinson, Atmospheric oxidation, in *Handbook of Property Estimation Methods for Chemicals: Environmental and Health Sciences*, R.S. Boethling and D. Mackay (eds), CRC Press, Boca Raton, 2000, 335–354 (Chapter 14).
24. P.H. Howard, Biodegradation, in *Handbook of Property Estimation Methods for Chemicals: Environmental and Health Sciences*, R.S. Boethling and D. Mackay (eds), CRC Press, Boca Raton, 2000, 281–310 (Chapter 12).
25. ECETOC, *Persistence of Chemicals in the Environment*, European Centre for Ecotoxicology and Toxicology of Chemicals, Technical Report No. 90, Brussels, Belgium, 2003.
26. J.S. Jaworska, R.S. Boethling and P.H. Howard, Recent developments in broadly applicable structure–biodegradability relationships, *Environ. Toxicol. Chem.*, 2003, **22**, 1710–1723.
27. T. Mill, Photoreactions in surface waters, in *Handbook of Property Estimation Methods for Chemicals: Environmental and Health Sciences*, R.S. Boethling and D. Mackay (eds), CRC Press, Boca Raton, 2000, 355–382, (Chapter 15).
28. D.G. Crosby, *Environmental Toxicology and Chemistry*, Oxford University Press, New York, 1998, 336.
29. P.G. Tratnyek and D.L. Macalady, Oxidation–reduction reactions in the aquatic environment, in *Handbook of Property Estimation Methods for Chemicals: Environmental and Health Sciences*, R.S. Boethling and D. Mackay (eds), CRC Press, Boca Raton, 2000, 383–418 (Chapter 16).
30. Environment Canada, *Toxic Substances Management Policy: Persistence and Bioaccumulation Criteria*, En 40-499/2-1995E, Final Report, Ottawa, ON, Canada, 1995.
31. E. Webster, C.E. Cowan-Ellsberry and L.S. McCarty, Putting science into PBT evaluations, *Environ. Toxicol. Chem.*, 2004, **23**, 2473–2482
32. E. Webster, D. Mackay and F. Wania, Evaluating environmental persistence, *Environ. Toxicol. Chem.*, 1998, **17**, 2148–2158.
33. M. Stroebe, M. Scheringer and K. Hungerbühler, Measures of overall persistence and the temporal remote state, *Environ. Sci. Technol.*, 2004, **38** (21), 5665–5673.
34. K. Fenner, M. Scheringer and K. Hungerbühler, Persistence of parent compounds and transformation products in a level IV multimedia model, *Environ. Sci. Technol.*, 2000, **34**, 3809–3817.
35. UN/ECE, Executive body decision 1998/2 on information to be submitted and the procedure for adding substances to annexes i, ii, iii to the protocol on persistent

organic pollutants, United Nations Economic Commission for Europe, Geneva, Switzerland, 1998.
36. A. Leip, and G. Lammel, Indicators for persistence and long-range transport potential as derived from multicompartment chemistry–transport modeling, *Environ. Poll.*, 2004, **128**, 205–221.
37. F. Wania, D. Mackay, Y.F. Li, T.F. Bidleman and A.Strand, Global chemical fate of α-hexachlorocyclohexane. 1. Modification and evaluation of a global distribution model, *Environ. Toxicol. Chem.*, 1999, **18**, 1390–1399.
38. F. Wania, and D. Mackay, Global chemical fate of α-hexachlorocyclohexane. 2. Use of a global distribution model for mass balancing, source apportionment, and trend prediction, *Environ. Toxicol. Chem.*, 1999, **18**, 1400–1407.
39. M. MacLeod, D. Woodfine, D. Mackay, T. McKone, D. Bennett and R. Maddalena, BETR North America: A regionally segmented multimedia contaminant fate model for North America, ESPR – *Environ. Sci. Pollut. Res.*, 2001, **8**, 156–163.
40. F. Wania, Assessing the potential of persistent chemicals for long-range transport and accumulation in polar regions, *Environ. Sci. Technol.*, 2003, **37**, 1344–1351.
41. W.A.J. van Pul, F.A.A.M. de Leeuw, J.A. van Jaarsveld, M.A. van der Gaag and C.J. Sliggers, The potential for long-range transboundary atmospheric transport, *Chemosphere*, 1998, **37**, 113–141.
42. D.H. Bennett, W.E. Kastenberg and T.E. McKone, General formulation of characteristic time for persistent chemicals in a multimedia environment, *Environ. Sci. Technol.*, 1999, **33**, 503–509.
43. A. Beyer, D. Mackay, M. Matthies, F. Wania and E.Webster, Assessing long-range transport potential of persistent organic pollutants, *Environ. Sci. Tech.*, 2000, **34**, 699–703.
44. M. Scheringer, Persistence and spatial range as endpoints of an exposure-based assessment of organic chemicals, *Environ. Sci. Technol.*, 1996, **30**(5), 1652–1659.
45. M. Scheringer, Characterization of the environmental distribution behaviour of organic chemicals by means of persistence and spatial range, *Environ. Sci. Technol.*, 1997, **31**(10), 2891–2897.
46. M. Scheringer, F. Wegmann, K. Fenner and K. Hungerbühler, Investigation of the cold condensation of persistent organic pollutants with a global multimedia fate model, *Environ. Sci. Technol.*, 2000, **34**, 1842–1850.
47. K. Fenner, M. Scheringer, M. MacLeod, M. Matthies, T. McKone, M. Stroebe, A. Beyer, M. Bonnell, A.C. Le Gall, J. Klasmeier, D. Mackay, D.Van De Meent, D. Pennington, B. Scharenberg, N. Suzuki and F. Wania, Comparing estimates of persistence and long-range transport potential among multimedia models, *Environ. Sci. Technol.*, 2005, **39**, 1932–1942.
48. D.C.G. Muir, C. Teixeira and F. Wania, Empirical and modeling evidence of regional atmospheric transport of current-use pesticides, *Environ. Toxicol. Chem.*, 2004, **23**(10), 2421–2432.
49. D. Mackay, W.Y. Shiu and K.C. Ma, *Physical–Chemical Properties and Environmental Fate and Degradation Handbook*, CRCnetBASE 2000, Chapman & Hall, CRCnetBASE, CRC Press, Boca Raton, FL, 2000. (CD-ROM.)

Subject Index

Acceptability, 48, 55
Acclimation, 112, 113, 124
Accumulate in biological systems, 16
Acid volatile sulfides (AVS), 116, 117, 118
Adaptation, 113, 124
Additives, 21
ADI, 69, 76, 77
Advisory Committee on Hazardous Substances, 16
Aerosol transport, 149
Air-water partition coefficient, 135
Alachlor, 149
Animals for testing, 10, 14, 17, 74
Antarctic, 147
Anthracene, 140
Arctic, 147
Asbestos, 35
Assessment endpoint, 92, 85, 91
Assessment factor, 88, 94
Assessment of risk, 7, 66, 68, 70, 71, 80
Asthma, 25
Atrazine, 149

BAF, 107, 108, 109, 124
Bahia Declaration, 15
Basel Convention, 15
BCF, 107, 108, 109, 124
Benchmark dose, 69
Best Available Techniques, 5
Bhopal, 36, 39
Bioaccumulation, 16, 66, 133
Bioaccumulation, 11, 106

Bioaccumulative properties, 12
Biocides, 34, 94, 96
Bioconcentration factor, 107, 124, 137
Biodegradation, 142
Bioinformatics, 73, 75
Biomagnification, 106, 107
Biomarkers, 70, 72, 73
Biotic ligand model, 115, 116, 120, 124, 125

Cadmium, 121, 123
Cancer, 24, 25, 29, 30
Car batteries, 27
Carcinogenicity, 66
Carrying capacity, 90, 91
Case-control studies, 71
Causality, 92
Characteristic travel distance, 149
Chemical assessment programme, 18
   equilibrium models, 120, 125
Chemical industry, 21, 23, 40
   intermediates, 38
   preservation, 22
   space plot, 140, 141
   Strategy on the Sustainable Production and Use of Chemicals, 15
   substitution, 10, 18
Chemicals, 13
   (Hazard Information and Packaging for Supply) Regulations, 6
Chlorine, 22
Classification and labelling of chemicals, 6, 7

Subject Index

CMR, 48, 54, 55, 56, 57, 66, 79
COC, 76
Cohort studies, 71
COM, 76
Complex effluent, 5
Concentration addition, 98
Confounder, 71, 72
Consents, 5
Contaminated land, 4
Copper, 121
Cost-benefit, 34
COT, 76, 77
Cross-sectional studies, 71

Dangerous Substances Directive
  Regulation, 6, 8
Data uncertainty, 47, 49, 51
DDT, 30, 107
Deterministic risk assessments, 88, 89
Detoxification, 106, 111, 113, 124, 125
DG SANCO, 45, 46,
Diet, 24, 116, 123
Diffusional limitation, 122
Direct toxicity assessment, 5, 25
Directive 2001/82/EC, 94
  2001/83/EC, 94
  2004/27/EC, 94
  98/8/EC, 94
Disposal of hazardous waste, 15
Dissolved organic matter, 108, 119
Dose-response, 27, 28, 68, 69, 70, 73, 76, 80
DYNAMEC, 11

Ecological risk assessment, 84, 102
Ecosystem goods, 89, 90, 91
  services, 89
Effect, 72, 73
  assessments, 86
Effective dose, 72
Effluent toxicity, 6
Endocrine disrupters, 31, 59, 80
  disrupting chemicals, 74
Endocrine disruption, 50, 55, 56, 67, 70, 80, 97, 98

Environmental effects, 25
  fate models, 114, 125
  media, 134
  quality, 3
  Quality Standard, 2
  risk assessment, 84, 86, 87, 89, 91, 94, 95, 99
  risks of chemicals, 16
EPAQS, 77
Epidemiological studies, 68, 71, 72, 76, 79, 80
Equilibrium models, 108
Essentiality, 106, 112
European Chemicals Agency, 11, 41
  Chemicals Bureau (ECB), 6, 132
European Inventory of Existing
  Commercial Chemical
  Substances, 8
  List of Notified Chemical
  Substances, 8
  Parliament, 41
  Union System for the Evaluation of
  Substances, 9
EUSES2, 50, 52, 53, 56, 60
Eutrophication, 25
Existing chemicals, 8, 93, 94, 96
  Substances Regulations, 8, 10, 16
Expert committee, 76
  Panel on Air Quality Standards
  (EPAQS), 76
Exposure, 26, 27, 65, 66, 66, 67, 68, 69, 70, 71, 72, 73, 74, 75, 76, 78, 79, 80
  assessment, 68, 76, 86

Fertilizers, 22
Food, 21, 118
  Standards Agency, 37
  web, 118, 126
Formulated mixtures, 21
  product, 38
Fractionation, 119, 120
Free-metal ion, 120
Freons, 133
Friends of the Earth, 36
Fugacity, 136, 137

Fulvic acid, 110
Functional redundancy principle, 90, 91

Gene, 73, 74, 75
Genotoxic carcinogenicity, 69
Geological background, 105, 124
Global warming, 25
Globally Harmonised System (GHS) of Classification and Labelling, 7
Greenpeace, 32, 36
Groundwater, 3, 4

Half-life criteria, 142, 143
Harmonised classification and labelling, 6
  Offshore Chemical Notification Format, 12
Hazard, 65, 66, 68, 78, 79, 104, 133
Hazard and risk assessment, 12, 17
Hazard assessment, 14, 27, 66, 133
  quotient, 104
Hazardous (special) waste, 13
  chemical, 1, 6, 11, 12, 15
Hazardous properties, 13, 17
  substances, 3, 4, 11, 12
Hazardous waste, 12, 13, 15
  Waste Regulations, 12
Hazards, 26, 38
HC5, 89
Health and Safety Executive, 7
  effects, 24, 25
Henry's Law constant, 135
Hexachlorocyclohexanes, 149
High production volume (HPV), 40, 133
Homeostasis, 107, 124, 112, 121
Human, 96, 97
  health, 1, 6, 10, 11
  medicines, 94, 95
Humic acid, 110
Hydrolysis, 143

ICCA, 40
Incineration, 71
Index dose, 70

Innovation, 34, 39
Insecticides, 31
Insurance hypothesis, 91
Integrated Pollution Control (IPC), 4
Intergovernmental Forum on Chemical Safety (IFCS) Bahia Declaration, 15
Internal dose biomarkers, 72
International activities on chemicals, 13
  Council of Chemical Associations, 13
  Uniform Chemical Information Database, 9

Knowledge uncertainty, 47, 50, 51, 52
Koch's postulates, 92

L(E)C50, 87, 88
Lead, 26
Lethal Dose 50, 28
Life expectancy, 22
Linearised multistage model (LMS), 70
LOAEL, 69
LOEC, 87
LOEL, 69, 76, 77
Long-range transport (LRT), 132, 133, 147

M&U Directive, 9
Marine environment, 2, 11, 12
Marketing and Use Directive, 9
Mass balance, 139
  transfer coefficients, 146
Mathematical models, 70, 73, 74, 75
Membrane, 106, 115
Mercury, 108
Metabolomic, 73, 75
Metal complexation, 110
  -binding properties of roots, 121
Metallothionein, 110, 111
Metals, 102, 105
Metolachlor, 149
Mixtures, 67, 70, 80, 97, 98, 99, 113, 125

Subject Index

Mode of action, 94, 98
Model uncertainty, 47, 76
Modelling, 57, 70, 76, 80
Models, 70, 75, 78, 114
Mode-of-entry, 146
Modes of action, 97
Molecular biological techniques, 74, 80
Multiple chemical exposures, 74
Mutagens, 66, 69, 78

Nanomaterials, 96
Nanotechnology, 59
Naphthalene, 140, 147
New chemicals, 8, 93, 94, 96
NOEC, 87, 88, 98, 99
NOEL, 69, 76, 77
Notification of New Substances Regulations 1993, 7

Octanol-air partition coefficient, 135
 -water partition coefficient, 106, 135
OECD, 8, 13, 40, 67
Oestrogen receptor (ER)-responsive genes, 74
Offshore activities, 12
 Chemical Regulations (OCR) 2002, 12
 Chemicals Notification Scheme (OCNS), 12
Organic food, 26
OSPAR, 11, 12

Paracelsus, 27
Partition coefficients, 132, 134, 135, 136, 137
PBT properties, 17
PBT, 10, 55, 56, 57, 60, 61, 79, 93, 94
PCBs, 107, 133
PEC, 86, 88, 94
PEC/PNEC, 87, 94, 95
Perception of risk, 66, 80
Perfluorinated organics, 149
Persistence, 11, 12, 16, 66, 133, 132, 142, 143

Persistent organic pollutants (POPS), 14
Pesticides, 30, 31
Pharmaceuticals, 31, 94, 96, 97, 98
Phenol, 140
Physiologically based pharmacokinetic (PBPK) model, 70, 73
Plant protection products, 85, 94, 95, 96
PNEC, 50, 53, 58, 59, 86, 88, 94, 95
Polar regions, 147
Pollution Prevention and Control (PPC) Regulations 2000, 4
 prevention principle, 90, 91
Polybrominated diphenyl ethers, 149
Poly-Parameter Linear Free Energy Relationships, 138
Pore water, 116, 119
Potato, 33
Precautionary approach, 15
 principle, 30
Preservation, 22
Prior Informed Consent (PIC), 14
Probabilistic risk assessments, 88, 89
Problem formulation, 85, 87
Prospective risk assessment, 85, 86
Protection goal, 89, 90, 91
Proteomic, 73, 75
Public's perception, 66
Pyrene, 140

QSAR, 57, 60, 61, 138
QSPR, 138, 142
Quantitative assessments, 73
 risk assessment, 68, 69
 structure-activity relationship (ASAR), 78, 80

REACH, 10, 11, 17, 34, 35, 38, 40, 42, 45, 54, 55, 56, 57, 59, 60, 61, 78, 79, 80, 93, 94, 96
Reducing discharges, 11
Reduction, 143
 strategy, 9
Reference dose (RfD), 69
Registration, 41, 93
Responsible Care, 39
Retrospective risk assessment, 85, 86

Rhizosphere, 123, 126
Risk, 65, 66, 67, 68, 69, 70, 72, 73, 74, 80, 104
  assessment, 7, 8, 9, 11, 12, 13, 26, 66, 68, 69, 70, 72, 73, 74, 75, 76, 78, 79, 80
  assessment model, 70
  characterisation, 46, 47, 53, 54, 58, 61, 68, 87, 96
  estimate, 68
  management, 66, 73
  perception, 66
  reduction measures, 9, 17
River basins, 3,4
Rotterdam Convention, 14
Route, 68
Royal Commission on Environmental Pollution, 29, 17, 32, 42

Scottish Environmental Protection Agency, 4, 5
Screening Information Data Set (SIDS), 13
Sediment, 116
Selection bias, 71
SIDS, 40
Silent Spring, 30
Soil solution, 119, 120
Solanidine, 33
Solubility products, 117
Sorption, 116
Speciation, 105, 108, 112, 119, 120, 124, 125
Spillover, 110
Stakeholder Forum, 16
Stockholm Convention of 2001, 14
Strategic Approach to International Chemicals Management, 14
Substitution, 23
Susceptibility, 73, 75, 76
Synergistic effects, 29
Synthetic organic chemicals, 32, 105, 107

TDI, 69, 76, 77
Technical Guidance Documents, 9
Testing methods, 6, 67, 78, 80
Toxic effect, 66, 67, 69
Toxicity, 11, 66, 67, 68, 72, 73, 74, 75, 78, 115, 116, 117, 118, 121, 133
  test, 67, 78
Toxicogenomic, 73, 74, 75
Toxicological methods, 74
  risk assessment, 75
  testing, 80
Toxiogenomics, 74
Transport velocities, 146
Transporter, 115
Trifluoroacetic acid, 133

UK Chemicals Stakeholder Forum, 16
  Chemicals Strategy, 15
  Committee on Toxicity and Chemicals in Food, Consumer Products and the Environment, 75
  New Substances Regulations, 10
  Offshore Chemicals Notification Scheme, 12
Uncertainties, 75, 76, 80, 87
United Nations Environment Programme (UNEP), 13, 14, 132
Uptake, 121

Vapour pressure, 136
Veterinary medicines, 94, 95, 96, 96, 97
Voluntary agreement, 1, 16, 18

Walden's rule, 136
Waste incinerators, 72
Water Framework Directive, 3, 92
  purification, 22
  Resources Act 1991, 5
WWF, 36

Z value, 137

Ollscoil na hÉireann, Gaillimh